Life Skills and Leadership for Engineers

Life Skills and Leadership for Engineers

David E. Goldberg
University of Illinois at Urbana-Champaign

McGraw-Hill, Inc.

New York   St. Louis   San Francisco   Auckland   Bogotá   Caracas
Lisbon   London   Madrid   Mexico City   Milan   Montreal   New Delhi
San Juan   Singapore   Sydney   Tokyo   Toronto

This book was set in Palatino by Better Graphics, Inc.
The editors were B. J. Clark and David A. Damstra;
the production supervisor was Paula Keller.
The cover was designed by Leon Bolognese.
R. R. Donnelley & Sons Company was printer and binder.

LIFE SKILLS AND LEADERSHIP FOR ENGINEERS

This book is printed on acid-free paper.

2 3 4 5 6 7 8 9 0 DOC DOC 9 0 9 8 7 6 5

ISBN 0-07-023689-5

Library of Congress Cataloging-in-Publication Data

Goldberg, David E. (David Edward), (date).
    Life skills and leadership for engineers / David E. Goldberg.—
1st ed.
        p.   cm.
    Includes index.
    ISBN 0-07-023689-5
    1. Engineers—Life skills guides.   2. Leadership.   I. Title.
TA157.G66   1995
620'.0023—dc20   94-5262

About the Author

DAVID E. GOLDBERG is professor of general engineering at the University of Illinois at Urbana-Champaign. He received his Ph.D. in civil engineering from the University of Michigan and has held faculty positions at the University of Alabama, Tuscaloosa and the University of Illinois. Prior to his doctoral studies, he held the dual position of project engineer and marketing manager at Stoner Associates, an engineering software and consulting firm in Carlisle, Pa. He received a 1985 NSF Presidential Young Investigator Award and is best known for his many discoveries, inventions, and publications in genetic algorithms—search procedures based on the mechanics of natural genetics and selection—especially his book, *Genetic Algorithms in Search, Optimization, and Machine Learning.*

For the Boybies, Max and Zack

Contents

Preface

This book owes its writing to my first engineering job out of school at what was then a very small engineering-software firm called Stoner Associates, Inc., in Carlisle, Pennsylvania. I remember thinking before I started how technically exciting it was going to be to work on four complex hydraulics computer programs. I remember wondering shortly thereafter if other engineers spent so much of their time writing, talking, selling, meeting, and always feeling too busy. In retrospect, perhaps my experience was a little extreme, but as I've talked with my own former students after they have started their first jobs, I've concluded that the average engineer walking out of school is surprised by the amount of nontechnical work he or she faces, by the amount of personal contact, the number of phone calls, meetings, reports, and presentations.

Now this is not to say that the new engineer is completely unprepared for these activities. Years of general education are supposed to—and no doubt do—instill some level of communication and people skills, but the hypothesis that this training is largely inadequate is supported by the number of leadership, management, writing, talking, and selling seminars that exist out in the real world to remedy a lack of practical capability. Wouldn't it be nice if today's engineering students received some intense training—a boot camp for the real world—through a practical examination of the life skills they will need in the first days, weeks, months, and years on the job?

This book attempts to provide a slice of that kind of practical training in a form that may be used in a classroom setting or for self-study. The book has been designed for a variety of uses: (1) in a one-credit course on leadership and life skills to be given to undergraduate engineering students a year or so before they graduate, (2) as a supplement to an introductory engineering class for underclass students, (3) as a supplement to a three- or four-credit senior capstone course, (4) as a supplement to a graduate seminar, (5) as part of course materials for corporate training or continuing education for new engineers, and (6) for self-study by engineering students and recent graduates.

The material has been tested in a one-credit senior seminar in General Engineering at the University of Illinois at Champaign-Urbana, but despite this experience, the teaching of these topics continues to be controversial. That the teaching of professional skills in an engineering curriculum might be controversial should come as little surprise if we look at the century-long tension between our profession's creative, technical, and business sides. Since World War II, the technical side—especially science and analysis—has been gaining ground, particularly in academic circles, but there is growing awareness that the more integrated design, manufacturing, and marketing processes embodied in concurrent engineering methods require engineers who can work in teams and who understand the needs of co-workers, customers, and stockholders alike.

In retrospect, perhaps the debate *between* design, analysis, and business was misframed and might better be resolved by recognizing engineering as the balanced combination of those skills that it is. In some ways, engineering is a gloriously marginal profession, but I use the word "marginal" without its usual negative connotation. Ours is an age of technology, and it seems that engineers can be— should be—among the leaders of this new age, but for this to come about fully requires the training of engineers with a balanced view of the importance of technical and nontechnical skills alike.

Therefore I hope the publication of this text will spur more engineering degree programs to install leadership–life skills training, not only because it will help more engineers enjoy more successful careers—and I believe it will do that—but because it is a matter more urgent than any one individual's success. There was once a time when industry was ruled by inventive engineering-manufacturing giants, titans like Thomas Edison and Henry Ford. In many parts of the economy, financial shenanigans and legal maneuvering have replaced product engineering and production activity as the surest road to the corporate suite. Although it is interesting that some of the most vibrant sectors of the economy—electronics, computers, and software, for example—are still run by engineers with bold product visions and enough people skills to make a company

DILBERT reprinted by permission of UFS, Inc.

happen, it is not encouraging that large portions of the economy have been taken over by people who have little interest in what they produce. The vision of this book is that this tide can be turned, that engineers with technical and people skills can help restore a passion for products and services to our boardrooms, to our design offices, and to our factory floors. It may or may not be an achievable goal, but we'll never know unless we try. The results of not trying are becoming increasingly unacceptable to a way of life we have valued but are now at risk of losing.

Acknowledgments

Many people provided much important feedback during the preparation of the early drafts of this manuscript. I thank Bob Barfield (Alabama), Larry Bergman (Illinois), David Brown (Alabama), Jim Carnahan (Illinois), Harry Cook (Illinois), Chuck Crider (Colonial Pipeline), Sydney Cromwell (Illinois), Kalyanmoy Deb (IIT Kanpur), Pete DeLisle (Illinois), Jerry Dobrovolny (Illinois), Chuck Evces (Alabama), Anju Jaggi (Illinois), Rachel Janssen (Illinois), Don Jones (General Motors), Ed Kuznetsov (Illinois), Angie Locascio (Illinois), James McDonough (Cincinnati), Dan Metz (Illinois), Dave O'Bryant (Illinois), Constantine Papadakis (Cincinnati), Carolyn Reed (Illinois), Tom Richwine (Stoner Associates), Kate Sherwood (Consultant), Angus Simpson (Adelaide), Mark Spong (Illinois), Mark Strauss (Illinois), Christy Tidd (Illinois), Debbie Thurston (Illinois), and Ben Wylie (Michigan) for their assistance. I also thank John Biddle (California Polytechnic–Pomona), and Dermot Collins (University of Louisville) for their helpful reviews of the manuscript.

I thank my wife, Mary Ann, for her support, especially during the difficult period when the original manuscript was drafted. This is the second book she has helped with (suffered through), and once again, she separated what was appropriate from what was simply off the wall. Finally, having children is one of the greatest lessons in life skills a person can have, and I thank my sons Max and Zack for teaching me a bushel of stuff I didn't know about or think about until they came along. For this reason I dedicate the book to them.

David E. Goldberg

Life Skills and Leadership for Engineers

CHAPTER 1

Life Skills and Leadership: Who Needs Them?

This book addresses two kinds of readers: newly practicing engineers and engineering "wanna-bes"—engineering students who are anxious to receive their degrees. If you're a practicing engineer, you'll probably find that what I'm about to say has the ring of truth. If you're a student, you may be in for something of a shock. You've studied many long years, and soon you'll plunge into that real world everyone has been talking about. Surprise! We—your engineering professors, your erstwhile advisors, mentors, and terrorizors—lied to you. We didn't do it intentionally, and we didn't do it explicitly, but we did it nonetheless. How? While you were spending 80 percent of your time studying difficult technical subjects—science, mathematics, and engineering—we forgot to mention that, though those subjects are important, they may constitute less than 20 percent of your working day and have less to do with your career success as an engineer than do your abilities to communicate with your co-workers, to sell your ideas, and to manage your time, yourself, and others. These crucial skills, these *life skills*, will determine your career success—and your happiness—more often than will your ability to manipulate a Laplace transform or analyze a statically indeterminate structure.

Of course, this is not to belittle these latter skills; after all, you wouldn't be an engineer unless you knew a great deal about technical matters. In a sense, the "lies" we told you were necessary to bootstrap you into a world of increasingly complex and changing technology. And don't get me wrong on this point either; even though student engineers spend most of their time learning technical stuff, they still don't know enough of it. New engineers will have to put in many hours listening to and learning from people who have been in the technical trenches. On the other hand, preparing for the difficult organizational and people-related challenges will help new engineers hit the ground running. Although I've addressed these words to the wanna-bes, the old hands know what I'm saying all too well. Life skills improvement is beneficial to engineers of all ages.

In a moment, I will list the topics covered in this book, but before I do, let me say what this book is *not*. First, it is not going to be a cookbook full of magic recipes for wealth, fame, and eternal life. There are already too many books of that genre that promise one or more of these things, and there is no need to chop down more trees in such wasted effort. Second, this book is not going to advocate any easy road to success through superficial or cosmetic changes. Although there are books out there recommending that you dress for success, lunch for success, hold a fork properly for success, and so on, I believe that nothing can substitute for the development of personal and organizational skills. I also know that engineers, as a group, are relatively unimpressed with superficial appearances and are more concerned with core competence, whether that competence is with things technical or with people. Of course, I'm not recommending that you dress shabbily, use improper etiquette, or any such thing. I'm only saying that we will concentrate here on developing key disciplines in managing yourself and working with others; I will leave your taste in neckties to you and your haberdasher.

This book and any course based on it will concentrate on tough issues—issues beneath the surface. This material will prove to be no cakewalk. For some reason, engineers and engineering students tend to think that things nontechnical are soft or easy. I can't speak to the content or rigor of nontechnical courses in academe, but the human side of engineering—the human side of life—is extraordinarily challenging. Although many of the things to be discussed here come under the category of common sense or conventional wisdom, it is true that common sense is sadly uncommon and that much of what passes for conventional wisdom is more conventional than wise. Therefore, just because we are going to concentrate on nontechnical skills for a time does not mean the going will be smooth or easy. Many of the things discussed will *sound* easy, but actually putting them into practice consistently and regularly will be among the toughest challenges you face, a challenge made even more difficult and frustrating because these things seem like they should be so straightforward to apply. They are not, and you should be prepared to come to terms with just who you are, how hard and smart you are willing and able to work, and what precisely motivates you to do your best.

This book is also going to steer clear of many of the more mundane aspects of living, some of which receive much ink in the literature of success. Yes, there are taxes to pay, investments to make, and insurance policies to buy, but these look less like opportunities than necessary evils. A key theme of this book is to become *engaged* with things we enjoy doing. This doesn't sound profound, but a lot of people spend a lot of time in conventional pursuits (like figuring taxes and making investments), never considering whether those activities bring them happiness or whether that time might be better invested in activities that do. Later we will return to this issue, but at the moment we need to outline more of what this book is actually going to cover.

SEVEN TOPICS

Having explained what the book is not, perhaps it is high time to say a word about what it is. This book, as already noted, concentrates on the life and leadership skills necessary today for success as an engineer. Specifically, it covers seven topics:

1. Writing
2. Presenting
3. Human relations
4. Time management and personal organization
5. Wealth and work
6. Brainstorming
7. Organizations and leadership

To cover such a broad array of topics in such a short book is a daunting—some would say, impossible—task. To give some chance of success the approach adopted is practical. The book is largely unconcerned with the theoretical or academic basis of what amount to time-tested solutions. We concentrate here on that which works. Of course, this should be a pretty good match with the nuts-and-bolts attitude of engineering practitioners and students alike. Moreover, the book tries to cut through the complexity of each topic by concentrating on a few key ideas. Such brevity risks increasing the possibility of misunderstanding or disagreement; however, even in these circumstances the book succeeds if those misunderstandings or disagreements challenge you to reexamine your approach to professional life.

In looking back over the list of topics, we can make a number of distinctions. Roughly speaking, the list divides into three categories: (1) communications, (2) resource management, and (3) groups and organizations. Writing and presenting are primary communication skills and we address them first and foremost. Graphical communication and visualization are also important to the engineer, but these topics are usually taken up as part of an engineer's technical training and are not considered here. Resource management topics are taken up at the personal level; they include the management of people, time, and money. The organizational topics consider why and how we work in groups.

THREE PRINCIPLES

Our examination of the seven topics will bring out three recurring principles. I have already spoken briefly of *engagement*—finding and doing things you enjoy profoundly—and although we'll not focus on engagement until Chap. 6, a number of the book's exercises will ask you to explore the ways in which your plans match those activities you find engaging. Fortunately, our profession is sufficiently multifaceted that it can comfortably accommodate a wide range of motivations and personal preferences.

A second principle that reverberates throughout these pages is that of *creating first, criticizing later*. In any creative activity, whether it is designing, writing, presenting, or problem solving, it is important to get many ideas out on the table to permit their cross-fertilization and to stimulate additional associations. School learning, with its emphasis on *convergent thinking*—applying what has already been discovered—discourages the *divergent* or *lateral* thought required for excellence in all creative activities. This theme appears most strongly in the chapters on writing and brainstorming, but it is general in its scope of application.

The last theme running through our topics is an *orientation toward others*. The economist sometimes has the luxury of imagining a "Robinson Crusoe" economy where a single individual's wants and needs are met by his or her own efforts, but almost all human activity is carried out through a messy mix of cooperation and conflict with others. Efficiency requires that we maximize the former and minimize the latter; to do this, we must understand the motivations of the other individuals involved. It is interesting that we are pushed in this direction not by altruism but by pursuit of personal effectiveness. It is also interesting that our study of others often results in our becoming better—marginally more objective—observers of our own behavior.

THREE ALARMS

The seven topics and three principles are important and we should hardly wait another minute to get started, but before examining the first skill—writing—I should warn you to use this book carefully. Specifically, three cautions must be exercised.

Be realistic in the application of ideals. Many of the skills we examine are discussed in terms of ideals. Doing this has benefits and risks. The primary benefit of using ideals is that we can easily define a target by which we can measure our own behavior and make adjustments. The primary risk is that no one can live up to ideals all the time. My introduction to the literature of success, which came during my first job, led to frustration; I read many books about business and management and saw that my company and my clients' organizations were far from the ideals discussed. Focusing on these discrepancies led to unproductive rounds of playing "ain't it awful." This is a rookie's mistake. One should be sensibly realistic in applying ideals. It may be all right to press for your own top performance or that of an organization that reports to you, but applying ideals to others who do not share your vision is a prescription for unhappiness and disappointment. Moreover, as pointed out elsewhere

(Fritz, 1991), you should select ideals with considerable care, as it is possible to become paralyzed by conflict between what the ideals promise and what is actually possible.

Appreciate, practice, and continually improve hard-won life skills. The second caution comes from an interesting conversation I had with a graduating senior after he had sat through my lectures at the University of Illinois. He told me that, after giving it some thought, he believed that there was some value in my lectures though at first he had found it very hard to focus on their message. He suggested that students have a hard time paying attention to an engineering professor's ideas about life skills, stuff that seemed like so much common sense. Needless to say, these weren't welcome words, but after giving them some thought, I decided that the student's reaction was a fairly natural one. All topics discussed in class (and all the things we discuss in this book) are "common sense" in the sense that we all must confront these issues at one time or another; however, just because a person has a basic grasp of a life skill doesn't mean that person has perfected his or her abilities. In talking about another complex skill—playing golf—champion golfer Ben Hogan wrote the following:

> It may seen that we have gone into unwarranted detail about the elements of the correct grip [of a golf club]. This is anything but the case. Too often in golf, players mistake the generality for the detail. They think, for example, that overlapping the finger is the detail and so they do not pay sufficient attention to how they do it. Or they confuse an effect (which can be quite superficial) with the action (the real thing) that causes the effect.*

Like golf, each of the skills to be discussed here is sufficiently complex that there are many subtle details or facets to learn; it is impossible to learn them in a single lesson. Moreover, it takes conscientious practice to maintain a skill once it has been developed. Just as golf professionals return over and over to the fundamentals of grip, stance, swing plane, and so on, so too must we return repeatedly to the basic skills and principles that help make us effective in our professional lives. And just as professional athletes take a positive attitude toward continual improvement—a philosophy the Japanese

---

*B. Hogan, with H. W. Wind, *Five lessons: The modern fundamentals of golf.* New York: Barnes, 1957, p. 30.

call *kaizen*—it is especially important for engineers, particularly new engineers, to adopt a philosophy of continuing professional development.

Engage the material and put it into practice. Sometimes when I look out at a classroom of engineering students during a lecture, I notice that when I put something on the board they often write it down; in fact, when I put an equation on the board they *always* write it down. But when I give a life skills lecture, I don't see much note taking. In fact, when I look out over a class, I see a mix of two types of faces. One type of face is that of the student who is really listening, and although I would feel better if those students were taking notes, I don't worry about them as much as I do about the other students, those with blank faces. I stop and remind these other students of some sage advice, "Life is not a spectator sport."

You won't agree with me on everything, but when you do, why not try applying those ideas to your work? When you don't agree with me, why not read what others have to say about the same issues? The bottom line in this book is that there is more to engineering than technical skill, and engaging the material herein, questioning it, and putting it to practice is a good start toward becoming a more effective engineer.

For Further Reading

BLOTNICK, S. (1980). *How to get rich your own way.* New York: Playboy Paperbacks.

CARNEGIE, D. (1981). *How to win friends and influence people* (rev. ed.). New York: Pocket Books.

DE BONO, E. (1973). *Lateral thinking: Creativity step by step.* New York: Penguin.

FRITZ, R. (1991). *Creating.* New York: Fawcett Columbine.

SCHELL, E. H. (1952). *The million dollar lecture and letters to former students.* New York: McGraw-Hill.

DILBERT reprinted by permission of UFS, Inc.

CHAPTER 2

Write for Your Life

I'm going to hazard a guess: I bet that you don't like to write. Sure, you can do it if you have to, and once you get going you're probably not bad at it, but writing is hardly your favorite activity. That's unfortunate, because writing is one of the most important of the life skills: It's a skill that can help you communicate your ideas and persuade others, a skill that can improve your personal productivity in every facet of professional endeavor. Moreover, and perhaps more importantly, writing can become an integral part of your thought process, enabling you to be a more consistent thinker by allowing you to carry on a debate with yourself in the only way that forces you to confront what you know and what you don't in a forthright manner.

In this chapter, we're going to examine why writing isn't a better-loved skill. Specifically, we'll examine the reasons why so many of us have mixed feelings about writing. Then, to counteract the fundamental problem, we'll adopt a useful writing exercise and a three-part writing process to help unlock the writer's block too many of us feel too much of the time.

WHY MANY PEOPLE DON'T LIKE TO WRITE

It's odd that so many thoughtful people dislike writing. After all, it is one of the few school activities that follow us from grade to grade and class to class, from the first year of elementary school to the senior year in college. How is it that such an important activity, one we have practiced and practiced, becomes so frequently blocked during our professional lives and is therefore so often dreaded or avoided?

The answer is simple and may be summed up in a single word: criticism. From our earliest school days, writing is an activity associated with making mistakes and receiving subsequent criticism from our teachers. Whenever I think of my second-grade teacher, Mrs. Brown, I can still hear her scold: "That's not a sentence," "Don't use a comma there," and "Watch your spelling." Because writing is such a complex activity, requiring the coordination of so many technical and creative skills, it does require a lot of feedback—much of it negative—for the student to master its basic technical aspects: grammar, spelling, punctuation, and sentence structure. Unfortunately, many of the people providing this feedback are less than supportive in correcting mistakes; worse, they are often unaware that constant criticism can damage the creative impulse. As we grow, we like to think that we have been able to cast off the yoke of such criticism, the emotional sting of these minor traumas, but more often than not they chase us into adulthood. In the case of writing, the ghosts of the Mesdames Brown of the world haunt us as we sit, pen poised over paper, immobile, not able to dash off so much as a phrase. And if we do dare to write, we are scared witless: We are afraid of making mistakes. How many times have you sat with a pen in your hand and ideas in your head, staring at a piece of blank paper? Then, when a sentence does finally pop out and onto paper, how often have you thought, "No, that's not right," and scratched the sentence out? Of course, this hasn't happened to you just once. Over and over, sentences pop out only to be cut down by Mrs. Brown's goblins: "That spelling's incorrect" and "Your structure's too simplistic." With a head full of evil spirits such as these, how can a person write? It's a lose-lose proposition if ever there was one: You're frustrated when you don't write because the job needs doing, but you're frustrated when you do write because the words won't come out just right. As a result, it is quite natural to avoid—or put off—an activity that prompts such a mixture of conflicted, largely negative thoughts; under such circumstances, there is little hope of ever enjoying the writing process.

FREEDOM FROM THE GHOST OF MRS. BROWN

Having identified the fear of criticism that blocks many writers, we are in a position to consider a straightforward remedy. The key to exorcising the ghosts and goblins of criticism is to *separate writing from revision*. The "write, cross out, write, cross out" mode described previously is the result of trying to do too much at once. We try to revise our work at the same time that we create it. These activities are distinct; some psychologists would even argue that they are processed in different hemispheres of the brain. Whether or not that is so isn't important here, but keeping separate these two very different activities is.

Separating writing from revision is a sensible suggestion—so sensible that I wish I could take credit for thinking of it, but I learned this valuable lesson from a book called *Writing with Power* by Peter Elbow. At the time I read that book, I was trying to confront the largest writing assignment I had ever encountered—my dissertation—and I was dreading the very thought of approaching something so long and complicated. I was browsing bookstores, looking for hints, clues, anything that would help me become a more confident and productive writer. From the first reading, Elbow's diagnosis and prescription struck me as just right. Of course, saying we should separate writing from revision and actually doing it are two different things. In much of the remainder of this chapter we will examine some techniques to help do just that. The first technique is a useful exercise called *freewriting*.

FREEWRITING

To eliminate the damaging legacy of criticism we must first and foremost learn to get our thoughts—whatever they may be—down on a piece of paper. Peter Elbow recommends a straightforward exercise for learning to do that. He calls it *freewriting*; as the name implies, it is writing that is not directed at a particular subject, project, or piece. Rather, it is simply an attempt to transfer thoughts to paper in an unfettered manner. To nuts-and-bolts engineers, the idea of writing without clear purpose may seem a little too "touchy-feely" to be of practical use. In a moment, we'll explore the rationale behind the exercise, but for right now we'll concentrate on its mechanics.

Freewriting is easy to do, the only physical requirements being a pad of lined paper and a pen. Thereafter, the process may be defined by a set of six rules:

1. Write for a predetermined, fixed period of time.
2. Do not stop and, insofar as it is possible, do not lift pen from paper until the time expires.
3. Do not cross out any writing, not one word.
4. Do not worry about spelling, grammar, punctuation, or structure.
5. Do not fix your mind on a particular topic, although if you remain on one topic that's all right.
6. If you become stuck, keep writing by repeating the same sentence over and over or by writing about how it feels to be stuck.

The rules are straightforward, and the best way to illustrate them is for me to stop talking and for you to start freewriting. So that you will be as uninhibited as possible, please realize that what you write during freewriting will be for your eyes only. You will not be asked to turn it in, nor will you be asked to share what you have written with others. In short, this is an exercise for you to learn about you. As a final reminder, don't stop, don't cross out, don't worry about grammar or usage, don't worry about topic, and if you get stuck, just write *something*. With these warnings, we are ready for the first of a number of so-called *online exercises*. If you are using this book as part of a course, your instructor will probably assign one or more of these exercises. If you are using the book for self-study, I strongly urge you to do each of the online exercises to get the most out of the book. With that said, let the freewriting begin.

Online Exercise 2.1

Freewrite for a timed session of six minutes. Follow the six rules of freewriting with almost religious zeal.

How did that feel? I can recall my first encounter with freewriting. What a strange experience it was! I remember having trouble restraining myself from crossing out, but soon the words began to flow. I also recall being surprised by some of the thoughts that started to surface, but most of all I remember the liberating feeling of being "allowed" to let different associated ideas flow one from the other. I'm sure your experience with, and feelings about, freewriting will differ, but the exercise is designed to do a number of things.

First, it is designed to let you experience what uninterrupted writing *feels* like. After all, for many people that feeling is uncommon enough that it is worth a special effort to achieve it.

Second, the rules of freewriting force you to view the words flowing from your pen less critically, thereby letting you generate more writing per unit time. How much did you generate in your first six-minute exercise? Many first-time freewriters generate between one-half page and one and one-half pages (single-spaced) in six minutes. Although freewriting is not directed at a particular task, it does give you some quantitative feeling for your potential to generate lots of material in a short time. To extrapolate your experience, simply multiply your six-minute page count by 10 to estimate your freewriting productivity on an hourly basis. For the average first-time freewriter this translates to something like 5 to 15 pages per hour. Even allowing for large amounts of wasted material, a writing process that allows you to move forward in a relatively uninhibited way has the potential for greatly increasing your writing productivity.

Third, and perhaps most important, freewriting is designed to give your writing a qualitative boost. Released from the bonds of judgment, you are able to generate new notions, partially formed ideas that previously would have been stillborn because they didn't come out perfect in wording, grammar, and usage. Thus, stripping away the tendency to be so self-critical can result in a step change in quality, the exploration of new rhetorical devices, and the discovery of what is often called a writer's *voice*.

Of course, all these things are possible, but one session does not a freewriter make. A sensible way toward improvement is to buy a notebook dedicated to your freewriting activities and to simply sit down at a designated time every day for a month, or two, or three, or six, and just freewrite. If you can get into the habit over a prolonged period of time, you will be able to follow the ebb and flow of your thinking and to play with words in a constructive way with no downside risk of criticism and with much upside potential for becoming a better, more productive writer. As you do more freewriting, I encourage you to play with variants of the exercise. A particularly practical one is to turn your writing consciously toward a specific topic. Exercises along these lines will loosen your writing so you'll be better able to create some required piece of writing. As you turn away from freewriting and move toward more directed writing tasks, you'll find that you need to modify the technique somewhat to bring it down to earth, but many of the lessons of freewriting will transfer quite easily to the directed task. In the next section, we will examine specific modifications to the freewriting exercise, together with a number of other techniques to motivate an approach to directed writing that works.

DIRECTED WRITING FOR THE REAL WORLD

The undirected exercise of freewriting helps loosen constipated pens, but by itself it is less than helpful in getting a needed piece of writing done. For that, you need a plan of attack, an approach to writing that will let you finish that memo, that section, that chapter, or that report in a timely fashion. In this section, we take the lesson of freewriting, modify it so we can direct the course of idea generation, and use it in conjunction with a specific approach to revision. We will call this approach *directed writing* and although I draw on Elbow's ideas, the process presented here is a hybrid, a process that has worked for me, and a process designed to give busy engineers an everyday approach to getting important writing tasks done well and quickly.

Directed writing consists of three components:

1. Quickplanning
2. Directed creation
3. Cut-and-paste revision

We'll look at each of these in detail.

Quickplanning

You'll note that I've avoided any mention of the writing teacher's perennially favorite tool: the outline. This omission has been intentional; I hate to give outlining any role in this process, because it is so often misused. As Elbow points out, outlining is an attempt to figure out what you're thinking about before you think it, and as we've seen in freewriting, we are perfectly capable of getting our thoughts on paper without the rigid discipline of lists. The dangers in constraining yourself to a rigid outline are twofold.

First, an outline prevents exploration of ideas as you think them; because you've got an outline you will be tempted to follow it in serial fashion. Unfortunately (for outline-based writing processes), the human mind is not a serial computer. It works by association, and when an idea pops into your head, it is probably worth exploring that idea right then and there, lest you lose the thought forever.

Second, the outline often misses important interrelationships between entries, interrelationships that should determine the ultimate ordering of the material. But, once again, if you've begun with an outline, you will be tempted (for reasons of pride, inertia, or sloth) to follow that original plan.

Thus it is easy to conclude that a rigid outlining procedure can be as much of a hindrance to good writing as a help. On the other hand, many writers—myself included—feel uncomfortable writing without some sense of where they are going. That is why I suggest the adoption of a coarse outlining scheme I call *quickplanning*.

To get the right mental image of quickplanning, think of the freewriting process; now, instead of writing down complete thoughts or sentences, imagine hopping around from one idea to the next, writing down only enough to convey each thought. As a mechanical device, it is useful to place a bullet—a large dot—in front of each idea fragment. Just write the thoughts down; don't worry about the order in which you generate them, and don't worry if the categories overlap. After you've written down the main points or ideas, you can massage the list into a preliminary ordering if you like, but if a good ordering isn't obvious, any ordering will do at this stage. You simply want quickplanning to give you mental keys that will help the associative process of generating raw text. I emphasize that this initial stage of planning should not be labored or lengthy. If you miss important points, you can always come back later and fill them in. You are simply trying to prime the associative pump for the next stage of directed writing: directed creation.

You may begin to see that the process suggested here is not as neat or as orderly as the usual one-shot, outline-and-write process that is often adopted. Of course, there is little that is orderly or neat about the human mind itself. That is not a criticism of our thought processes, however. Our minds are rich association machines that carry our thoughts along a synaptic superhighway of concepts, ideas, and relationships. One of the jobs of an appropriate writing process is to key into the way our minds work and to help us explore various outposts of thought along the neuronal road.

Directed Creation

With a rough quickplan of key topics before you, it is a straightforward matter to generate raw material for your piece. Again, you want to achieve a frame of mind similar to that of freewriting, except now you want to take away the "free" and direct the writing toward the quickplan form of bulleted topics. You should be careful to maintain the exploratory spirit of freewriting and not restrict yourself too much. If directed creation generates promising associations, why not go with them? You can even afford to be distracted by the occasional tangent (if you have the time), because sometimes by writing through tangential thoughts, you can discover something important about the topics of primary interest, things that would not have surfaced otherwise.

If we keep these things in mind, we can define the process of directed creation by seven rules:

1. Write associatively, using the quickplan bulleted topics as a guide until your thoughts run dry or until the time budgeted for each topic runs out.
2. Do not stop and, insofar as it is possible, do not lift pen from paper.
3. Do not cross out any writing, not one word.
4. Do not worry about spelling, grammar, punctuation, or structure.
5. Stick to the point somewhat, allowing some time to develop related topics and tangents.
6. If you become stuck, move on to the next bulleted topic or shift to a thought generated elsewhere in the process of directed creation.
7. Write on every other line, and do not write on the back side of a page.

As you can see, these rules are similar to those of freewriting, with a number of modifications to the first six rules and the addition of one other.

The changes to rules 1 and 5 reflect the primary modification to freewriting necessary to achieve a more directed flow. Again, the bulleted topics developed during the quickplanning session should be used as mere seeds for the larger piece; in general, be more trusting of what's in your head, and follow your writing where it leads. If you are writing to a deadline, you can cut off tangents that don't seem germane and allocate fixed amounts of time to generate writing on a particular topic—but some freedom should still be given to pursue thoughts as they come.

Rules 2, 3, and 4 remain largely intact; it is particularly important to follow rules 2 and 3, to keep the writing moving forward without crossing out.

Rule 6 has been modified to recognize that indeed you are now writing about *something*. If you should get stuck (or run out of things to say about a particular topic), you can simply move on to the next bulleted item. Of course, when the spirit moves you, you should not hesitate to move back to a partially treated item.

Rule 7 has been added to allow for the cutting and pasting of snippets of writing during the process of revision. Rule 7 recognizes two things: (1) It is useful to have extra space in between lines so you can add missing material, and (2) it is inconvenient to paste one page to another when material is written on the back of the sheet being so affixed.

Together, the seven rules form a basis for directed creation. As before with freewriting, the proof is in the doing, which leads us directly to our second online exercise.

Online Exercise 2.2

Quickplan and perform directed creation for a two- to four-page, double-spaced, handwritten essay on the topic "Where Do I Want to Be in Five Years?"

The "where" here is not geographical. The question asks you to explore your goals and aspirations. In answering, try to be concrete. Generalities and platitudes are not helpful in exploring what you think, nor do they make for very lively writing. Also, don't spend much time apologizing for your lack of certainty. It is hard to project what will happen tomorrow, let alone what will happen in five years. On the other hand, a vision and goals are useful in evaluating opportunities and making decisions. Remember that it is unnecessary to force your writing to be too coherent at this point; you will have the opportunity to reorganize, interpolate, and extrapolate when you perform cut-and-paste revision in a later section.

An Aside: Writing by Hand or by Computer?

The widespread availability of word-processing software and personal computers raises important questions as to how to integrate these tools into one's writing. Although computers profoundly change the final *preparation* and *presentation* of a manuscript, they really have little to do with developing an effective *process* of writing. After all, the grayware between our ears is the most important contributor to an effective piece of writing, not the software we use for manuscript preparation and presentation or the hardware on our desks.

Having said this, I have to concede that in this day and age most manuscripts will be processed at some point on a computer. The appropriate question, then, is at what point should computers be introduced into the writing process? At one extreme, some might argue for introduction of word processing at the very start: Compose text directly at a computer keyboard. At the other extreme, some might argue for entry of text only following the completion of an edited manuscript. Here, I come down firmly for both extremes (and for some alternatives in the middle), depending upon the length and nature of the document and the level of experience and confidence of the writer.

Direct entry of text into a computer seems to offer the advantage of timesaving, but there are two dangers associated with direct composition at keyboard: nice-output syndrome and hurry-up-and-wait disease. *Nice-output syndrome* occurs when an inexperienced writer looks at the nicely formatted output that comes off the laser printer and hesitates to revise the writing. When words are neatly printed, there is a tendency to treat them with more respect than they deserve. After all, they are nicely printed and look pretty darn professional, don't they? But good writing involves a good bit of fooling around with the initial output. Anything that makes us hesitate to try different combinations or orderings is something to be avoided.

Hurry-up-and-wait disease occurs because a good touch typist can easily outrace the speed with which he or she can think (Knuth et al., 1989, p. 14):

> Upon receiving a question from the audience concerning how many times he actually rewrites something, Don told us (part of) his usual rewrite sequence:
>
> His first copy is written in pencil. Some people compose at a terminal, but Don says, "The speed at which I write by hand is almost perfectly synchronized with the speed at which I think. I type faster than I think so I have to stop, and that interrupts the flow."

In some ways, these suggestions for intentionally slowing down the writing process through the choice of a particular medium (for instance, pencil and paper) parallel the just-in-time methods of modern manufacturing that emphasize a steady line speed and well-synchronized operations. Note that the author of the passage quoted above is talking about fairly involved manuscripts, like journal articles and books, which are both novel and fairly long. In composing long, new material, many writers would do well to write raw text out in longhand to avoid the hurry-up-and-wait disease.

On the other hand, many business documents are neither novel nor long; instead, they are fairly short and on well-worn topics. In these cases, there is little risk in direct composition. For example, memos, letters, brief reports, and similar documents might not require much handwriting. Especially when you've written extensively about a subject, your thinking *can* keep up with your typing, so the hurry-up-and-wait disease is not as much of a concern. Also, with shorter documents, the amount of reordering possible is so limited as not to be a worry.

Having said these things, however, it is still important for all writers of documents, long or short, to have a good feel for the writing process. Therefore, for the exercises in this book, and for writers who are trying to improve, I recommend handwriting for text generation and physical cutting and pasting for revision. Even if you rarely actually use this mode of writing, the mental image of preparing your document in this way can only help you straighten out your logic and flow.

Cut-and-Paste Revision

Having practiced directed creation, you are now free to plow forward. Not constantly trying to get things right the first time, you should have little trouble generating fairly large numbers of pages of raw material. With this important capability under your belt, it is time to learn to put this raw material into a more organized form through cut-and-paste revision.

Cut-and-paste revision has a number of physical requirements:

1. A red pen
2. Scissors
3. A glue stick
4. Sheets of unruled paper

These items, together with a ruled writing pad and pen, form your cut-and-paste tool kit.

To actually begin the cut-and-paste process, first read back through the raw output of your directed-creation session, cutting out those passages that contain something important, whether that something is a word, a phrase, or a paragraph. Then treat the assembled fragments like a puzzle, attempting to fit the pieces into an orderly sequence.

Of course, some of what you've already written will itself need revision. That's exactly why you skipped every other line, leaving room for interpolation. Also, at this point, a red pen is your best weapon for slaying wordy mammoths that roam through your sentences and paragraphs.

As you piece the puzzle together, you'll want to make the draft more permanent by gluing the fragments to another sheet of paper. Blank xerographic paper works well for that purpose, but any letter-size sheet of paper will do. As you paste pieces together, you will notice awkward transitions and even whole subsections that need rewriting. There is no need to panic. Simply throw yourself into directed-creation mode, generating text to fill the current need; as you switch between modes, however, it is important to recognize the mode shift. Don't fall into the bad habit of trying to write any of the new material right the first time. Instead, when you click into directed-creation mode, obey all the rules of that game. When you've written enough stuff to fill in the missing blanks, switch back into revision mode, cutting, red-penning, and pasting to your heart's content.

Again, as we discussed earlier, depending upon the length and novelty of the document and the experience and confidence of the writer, it may be fine to do the cutting-and-pasting phase on a computer. Even so, it is good to have experienced *real* cutting and pasting, at least once. The sense of writing as experimentation comes about most easily when you actually *do* cut and paste little scraps of paper together. Having that mental image can be invaluable for the writer who prefers to compose and revise at the keyboard.

It has taken some time to describe the whole process, but now that we have, it's time to finish the exercise we started previously.

Online Exercise 2.3

Using the material generated in Exercise 2.2, perform cut-and-paste revision for the two- to four-page, double-spaced, handwritten essay, "Where Do I Want to Be in Five Years?" Do not type the final document. Simply edit and complete the cut-and-pasted copy by hand.

Having undergone the entire process once, you can begin to see the productivity gains and qualitative advantages of separating writing from revision. Of course, becoming a better writer is a never-ending quest; becoming a better editor is an important milestone along that road.

EDIFYING EDITING

Separating writing and revision through the use of directed writing and cut-and-paste revision can be such a liberating experience that the recently initiated may get the idea that the first products of these efforts need little or no additional attention. This idea is usually a mistaken one; most drafts need a number of passes of careful editing before they are ready to meet their final audience.

Mrs. Brown and company spent most of their classroom hours on matters of grammar, punctuation, sentence structure, and organization. They spent far less time in teaching matters of style. There are a number of editing maxims that can help enliven your writing:

1. Omit needless words.
2. Use active voice.
3. Enhance parallel structure.
4. Watch rhythms.

This list is far from exhaustive, but its guidelines are among the most important. At the end of this chapter, we'll list some books that give additional stylistic suggestions. Here we concentrate on considering each of our four guidelines in somewhat more depth.

In the absence of any other stylistic guidance, we can do much worse than to follow Will Strunk's advice and "omit needless words" (Strunk and White, 1979, p. 23). While it is something of an exaggeration to say that all good writing is brief, much bad prose is wrapped in layer upon layer of verbal fat, and the dietician's red pen is often in order. Good examples of vigorous and not-so-vigorous writing can be found in Strunk and White.

Active voice is essential to vigorous writing, and it is in this regard that so much technical writing falls flat on its face. Some of the stumbling is self-inflicted through adherence to the old saw "Never write in the first person." While it is true that it would be unwise to adopt the somewhat breezy style of this book in a typical consulting report, I doubt whether Western commerce would come to a screeching halt if the collective "we" or the individual "I" were used in more business writing. This single step immediately forces writing to become more active and lively. Even if you are required to write in the third person, it is possible to enliven your writing by avoiding anonymous subjects and talking about project teams, projects, or experiments more personally.

Pursuit of parallel structure is an important principle that is often ignored by the less-experienced writer. For example, suppose the listing of topics for this section had been written as follows:

1. Omit needless words.
2. Use active voice.
3. Enhance parallel structure.
4. Rhythms should be watched.

Each phrase is grammatically correct, but maxim 4 breaks the parallel structure set up in numbers 1 through 3. The verb-noun order established in the first three is upset by the noun-verb order of number 4. The change in pattern is mentally disruptive and ultimately prevents the writing from being as vigorous as it could be. Existing parallel structure should be identified and enhanced; building parallel constructs can bring an immediate power boost to much writing.

Attention to rhythm or meter is most often associated with poetry, but good cadence in prose is no vice. (Prose is just poetry that can earn a living.) In fact, many stylistic rules relate to an attempt to achieve better meter or rhythm in prose writing. For example, a common error of the inexperienced writer is to string together sentences of similar length and structure—the see-Spot-run, run-Spot-run syndrome. Although such sentences are grammatically correct, the error is one of boringly repetitious structure.

Although there is no foolproof way to recognize and correct bad cadence, it is useful to read your writing out loud. Once you identify passages with awkward meter, they sometimes can be fixed through the joining that comes with the use of a conjunction, the splitting that comes from the use of a semicolon, or the careful rearranging that achieves well-modulated structure. Though at times more major surgery is necessary, reading your writing aloud and listening to its rhythm can help improve its power and impact.

IMPROVING YOUR WRITING

Good writing is a journey, not a destination, and there are a number of side excursions worth the fare:

- Reading more
- Writing more
- Getting professional editorial help

We briefly examine each of these possibilities.

One of the easiest ways to improve one's writing is to read more, and one of the easiest ways to do that is to read a first-rate newspaper. If you don't already take a regular paper (and even if you do), consider *The Wall Street Journal*. It has some of the best writing of any major national newspaper. Moreover, the style that its editors have adopted—far from being the turgid business prose you might imagine if you've never read the *WSJ*—is some of the most lively and engaging newspaper writing you will find anywhere. With so many business leaders getting their daily fix of the *WSJ*, you can do much worse than to use that paper's prose as a model for your own improvement.

Good nontechnical, nonfiction writing can provide a model for exposition. Well-written history and biography (for example, works by Martin Gilbert, Paul Johnson, and William Manchester) are useful for learning to handle time and sequence. Well-written popular science books (by such authors as Isaac Asimov and James Gleick) are useful in learning that the presentation of factual information need not be dull.

Much has been written about writing, and engaging the literature of writing can be helpful, but it is important to pick books that ring your truth bell. I've already mentioned Peter Elbow and *Writing with Power*; this is a good starting place if you are interested in learning more about writing as a process. I've also referred to *The Elements of Style* by Strunk and White, an amazing little volume that's on almost everyone's short list of writing books. For matters of grammar and punctuation, I recommend Karen Gordon's *The Transitive Vampire* and *The Well-Tempered Sentence*. These clever books cover difficult material using quirky, humorous examples and counterexamples. For a fairly encyclopedic view of almost all matters of detail and form, there are few better sources than *The Chicago*

Manual of Style. Other recommendations are presented in the reading list at the end of the chapter.

Another easy way to improve your writing is simply to write more. Because writing is such a painful process for so many of us, we generally volunteer to write about as often as we might sign up for a root canal. Shortly after I joined the faculty at the University of Alabama, I was "volunteered" to be the secretary of the weekly faculty meetings. Rather than gripe, I took that opportunity to develop my writing skills (and occasionally to tickle my colleagues' funny bones), and along the way I learned a good bit about conveying mundane details in an interesting way. Similar opportunities pop up from time to time for everyone, whether at work, in a civic or church organization, or simply in writing letters to friends and relatives. The prevalence of electronic mail (E-mail) is also an opportunity to try your hand at informal pieces from time to time. I urge you to explore these and other writing opportunities.

If you have the opportunity to get professional editorial help, take it. Having a "pro" go through your work helps you identify your most common errors of grammar, punctuation, and usage. A good editor can also help with sentence, paragraph, and overall structure. English departments at universities and community colleges can often put you in touch with a qualified editor, but it's a good idea to get references from writers who have worked with the individual before making any commitments.

Summary

In this chapter, we've considered the process of writing. We've seen how the criticism of early efforts has left many professionals uncomfortable with their writing ability. We have considered an exercise called freewriting, which permits us to explore our thoughts without fear of criticism as we write with no particular purpose in mind. We have then seen how to bend the process of freewriting to a more directed piece and recognized how much directed writing followed by cut-and-paste revision can be a fairly effective means of completing the writing tasks before us. For many people, following these steps is enough to unlock the writer inside. The steps require practice, and at first it is difficult to avoid being critical during the creative phase of the work, but if you can discipline yourself to do so, the dividend can be nothing less than a lifetime of productive and confident writing.

Offline Exercises

1. Freewrite for 10 minutes every day for two weeks. In a short paragraph, compare and contrast the quantity and quality of your freewriting on the first and last days of the trial period.
2. Use the directed-writing process to write a brief essay, taking either the affirmative or negative position, on one of the following topics:
   a Engineers are properly appreciated in society.
   b Engineers do not know enough about nontechnical topics.
   c Laypeople do not know enough about technical topics.
   d America's competitive position is declining because American workers are lazy.
   e American executives are paid too much.
   f The United States is a nation in decline.
3. Write a brief essay arguing the opposite side of the issue you chose in Exercise 2.
4. Write a letter seeking employment with a company that interests you.
5. Write a thank-you letter to an individual who interviewed you for a job at a company.
6. Write a short autobiography.
7. Write a resume.
8. Write a short family history tracing some members of your family back at least two generations.
9. Write a brief biography of an interesting family member or an acquaintance.
10. Write a set of instructions for a game, a machine, a piece of software, or other complex entity used by human beings.
11. Write a brief description of the principles of operation of a device, process, or algorithm.
12. Write a brief paper describing the solution of a calculus or physics problem.
13. Write a brief how-to article on some aspect of a favorite sport or hobby.

14. Write an eyewitness account of the most unusual social gathering you ever attended.
15. Write copy for a brief marketing brochure for a technical product or service with which you are familiar.
16. Form a group of five writers and choose a topic from the list in Exercise 2. Have a brief group quickplanning session and then divide the bulleted items among the group members. Each member should then perform cut-and-paste revision using the raw material of the entire group. Exchange results; discuss the differences of organization and style represented by the five essays.
17. With another writer, select a topic from the list in Exercise 2. One of you should do directed creation for the affirmative, the other for the negative. After both of you have completed your initial writing, each writer should take *both* sets of raw material and generate a balanced essay that examines both sides of the chosen issue. Discuss similarities and differences in your final essays.

For Further Reading

Elbow, P. (1981). *Writing with power.* New York: Oxford University Press.

Gordon, K. E. (1984). *The transitive vampire.* New York: Times Books.

Gordon, K. E. (1983). *The well-tempered sentence.* New Haven, Conn.: Ticknor & Fields.

Hacker, D. (1989). *A writer's reference.* New York: Bedford Books.

Knuth, D. E., P. Larrabee, and P. M. Roberts (1989). *Mathematical writing.* Mathematical Association of America.

Mathes, J. C., and D. W. Stevenson (1991). *Designing technical reports* (2d ed.). New York: Wiley.

Safire, W. (1990). *Fumblerules: A lighthearted guide to grammar and good usage.* New York: Doubleday.

Safire, W., and L. Safir (eds.) (1992). *Good advice on writing: Writers past and present on how to write well.* New York: Simon and Schuster.

Strunk, W., Jr., and E. B. White (1979). *The elements of style* (3d. ed.). New York: Macmillan.

Trimble, J. R. (1975). *Writing with style.* Englewood Cliffs, N.J.: Prentice-Hall.

University of Chicago Press (1993). *The Chicago manual of style* (14thz ed.) Chicago, Ill.: Author.

CHAPTER 3

Power Presenting

It seems that the United States is always in the middle of an election. And elections mean speeches: stump speeches, impromptu speeches, TV speeches, speeches after dinner, speeches in schools, and speeches in malls. With the prevalence of stand-up public speaking in the political arena, it would be easy for the new engineer to become confused and assume that the same mode is commonplace in business communication. The truth is that it is the rare businessperson today who gives a stand-up speech, at least for everyday business matters. With the advent of overhead projectors and transparencies, businesspeople stopped giving speeches and started making *presentations*: talks supported by projected visual material. In this chapter, we'll examine the reasons why you should join this mass movement and learn to prepare and deliver effective, transparency-based presentations. Specifically, we'll examine the reasons why you should *present*—not speechify—and we'll consider the elements and the process of presentation preparation. We'll also consider methods for preparing effective transparencies and ways to sharpen your delivery skills.

WHY PRESENT?

After you have some experience giving transparency-based presentations, the reasons for presenting rather than speaking will seem almost self-evident. That this knowledge is not genetically transmitted was brought home to me several years ago when I was a group advisor for a GE 242 Senior Design Project at the University of Illinois. The design team—a talented group of motivated senior engineering students—had prepared their spiel and wanted to rehearse in front of me, so I could give them feedback. As soon as they began to talk, I realized—to my amazement and chagrin—that they were giving a speech. Sure, they had a transparency or two as window dressing, but it was a speech they delivered, with one student working from notecards, one reading verbatim from a script, and one working from (a faulty) memory. After listening for a few minutes, I asked them why they were making their lives so difficult. Uniformly, they answered that they thought that this was what the big guys did and that this was what they were expected to do for their first shot at the big time. Fortunately, they were persuaded that they could be more relaxed—and more communicative—if they would present, not speechify. They went on to give a solid, transparency-based presentation, but the experience left me with the impression that there is a wide gap between what students *think* a business presentation is and what such presentations *really* are.

Formal oral communication in business is now dominated by transparency-based presentations for two good reasons:

1. Presenting is easier to do.
2. Presenting conveys more information to your audience.

One of the reasons I couldn't believe my students chose to make a speech was because I know how hard it is to make a good one. Consider the three ways you can give a formal speech. You can memorize the speech—but for most people it is too easy to forget portions and become flustered. You can read from a script—but, unless you are a good actor, it is very hard to read and not sound as though you are reading. Or you can work from notecards (and this is perhaps the most sensible way for the occasional speech maker to work)—but even then, because your audience is not visually occupied, they will focus more on you, your mistakes, and your use of notecards. The presenter needs no such extra attention; nor does he or she need the stress it can cause. Contrast the difficulties of making a speech with the utter ease of giving a well-planned transparency-based presentation. You walk up to a projector with transparencies (large notecards made of transparent plastic with your notes printed in large, readable letters), and you fill in the blanks of your transparencies verbally as you slap one after another down onto the projector surface. While you're doing your thing, your audience isn't minding you much. They are happily engaged with the material you are putting before their eyes, the same material that is providing you with cues to continue your talk. In this way, presenting is much more forgiving to the communicator, providing notes and props to help get through the talk.

Presenting is also advantageous to the audience. Well-planned transparencies provide a second channel of information, augmenting the primary source—the speaker's voice. An audience member who misses a point from one source can often pick it up from the other. Moreover, in this video age it is not irrelevant that a transparency presentation is a visually engaging activity. An audience raised on television expects to have its visual field filled. A communicator who misses or misunderstands such an important audience expectation is simply asking for trouble.

Thus we are drawn to an interesting conclusion. When you have a choice, choose to present. You'll be more relaxed, the audience will be better engaged with the material, and more information will be conveyed.

PREPARATION MAKES THE PRESENTATION

That a presentation is easier to deliver than a speech certainly does suggest that we should choose to present whenever we have a choice, but ease of delivery does not imply that a presentation requires less preparation than a speech. In fact, because of the need for transparencies, a good presentation usually requires more time for preparation than a comparable speech. Over the next few sections, we'll consider the critical aspects of designing an effective presentation: audience analysis, subject selection, elements of a presentation, preparation process, and transparency design and preparation.

Audience Analysis

The best presentations come from serious consideration of audience. Experienced writers and speakers have a gut feel for their audiences and constantly adjust to audience needs, but the less-experienced communicator has less of a feel and has to give more consideration to the approach chosen. Fortunately, this requires no new, hard-won skills; the writing process of the previous chapter will be the primary tool we'll use, along with a list of audience characteristics. Together, these two tools will permit the writing of an *audience brief* to help guide presentation planning for a particular audience.

In preparing the brief, there are three audience characteristics to keep in mind:

1. Motivation
2. Patience level
3. Educational and technical background

By far the most important of these characteristics is audience motivation. Why is your audience bothering to sit there and listen to you? Are they simply scratching an intellectual itch, or do they need to learn something fairly specific? Are most audience members there for the same reasons, or are different audience members there for different reasons? It is important to address both the motives and the *variance* in motives as you contemplate your target audience. Of course, the main reason that you contemplate the audience is to connect your material to their motives. This permits you to establish the appropriate *angle* from which to present your subject matter. Any subject can be presented from various perspectives. For a relatively homogeneous audience you may select one particular point of view; for a more mixed audience it may be necessary to present multiple viewpoints to connect with differently motivated individuals. Only by understanding motives, and by focusing your subject material toward those motives, can you hope to reach your audience—and thereby accomplish your goals as a presenter.

Patience may be a virtue, but in business, time is money; many of your audience members will have severe limitations on how long they can sit still for your message. Therefore it is important to tailor the length of your talk to the level of patience (or, more often, impatience) of your audience. For example, a CEO with nine appointments before lunch and a plane to catch has one patience level, and a project engineer with only his workstations on hold has another. If you're faced with a situation in which both high-level managers and project engineers compose your audience, it may make sense to split your talk into distinct management and technical briefings, and this is often done. On the other hand, there will be times when you can expect fairly uniform patience levels in your audience—at technical society conferences, for example. Even then, audience patience can become an issue, especially if your talk exceeds the allotted time.

Just as levels of patience and length of talk must be matched, so must audience background and the intellectual level of a presentation. Waxing eloquent about a set of differential equations in front of an audience of Realtors is likely to evoke thoughts of cost variation among leases (differ-rental equations). Of course, spending time explaining elementary differential equations to a group of Ph.D. physicists is equally nonsensical. Therefore, assessing your audience's technical and educational background is vital. Once again, the trickiest audiences are those with mixed backgrounds. In the worst cases, it may be best to divide if one wants to conquer the heterogeneous audience.

Considering these aspects of your audience will help you design your presentation appropriately. Table 3.1 summarizes the connections between audience characteristics and presentation consequences. Although this section has been fairly clinical in dissecting the components of audience analysis, the more appropriate mechanism for considering a particular audience is the holistic preparation of an *audience brief*.

The next online exercise asks you to prepare an audience brief for a presentation that continues the theme of the exercises in the previous chapter. If this exercise is to be performed in class, you may

TABLE 3.1. Audience Characteristics and Presentation Consequences

Audience Characteristic	Presentation Consequence
Motivation	Angle
Patience level	Length
Education-technical background	Intellectual level

Online Exercise 3.1

In the course of this chapter, you will prepare a seven-transparency presentation on "Where Do I Want to Be in Five Years?" (You may change the title if you like.) For this exercise, write a one- to two-page, double-spaced, typewritten audience brief that identifies the motivations, patience level, and background of your target audience.

---

want to do an analysis of that target audience; if you are doing the exercise on your own, choose an imaginary audience (for example, parents, colleagues, or potential employers) and stick to that imagined audience throughout this chapter's online exercises.

Subject Selection

Subject selection will, for us, be something of a short subject. Often a business presentation arises out of a particular organizational need, making subject selection something of a moot point; however, there is some room even within the confines of a predetermined subject to choose among various aspects to be included in your presentation. There will be also be times when you are in greater control and can choose your subject fairly freely—for example, when you choose to make a presentation at a technical conference or before a civic or church group. In any event, the prime directive of subject selection may be stated quite simply:

> Within the constraints of organizational need and audience characteristics, choose a subject about which you are both *knowledgeable* and *enthusiastic.*

Perhaps this seems like simply stating the obvious, but how many times have you seen presenters choose to speak when they knew little about a topic for which they had not very much enthusiasm? Audiences can see through an amateur in a New York minute—if not during a talk, then certainly during the question-and-answer period. Audiences are also fairly savvy at detecting whether a speaker has enthusiasm for his or her topic. The best reason to choose a topic is your passion for that topic: Good presentations don't grow on trees, and your enthusiasm will carry you through as you plan, prepare, and deliver your talk.

Elements of a Presentation

In presenting, you can do worse than to follow the adage "Tell them what you're going to say, say it, then tell them what you said." A good presentation sets the stage in the listener's mind, presents the core material, and sums up and spells out the consequences of what was said.

A simple structure that accomplishes these things contains the following elements:

1. Title
2. Foreword
3. Overview
4. Body
5. Summary and conclusions

In the remainder of this section we'll examine each of these in some detail.

Title. It may be somewhat unusual to think of the title of a presentation as a separate element (and it is true that most titles are afterthoughts—more *after* than *thought*), but a good title can be critical to the success of a presentation. It is the first element that the audience sees or hears; it is important both for creating a positive first impression and for building in your audience the desire to hear more. To do these things, a good title should be *informative*, reflecting the material contained in the presentation, and *interest-provoking*, creating desire and anticipation.

Selecting titles is a somewhat mysterious art, and rather than be too analytical about it, let me give a sampler of actual presentation titles, ranging from the fairly straightforward to the more purely provocative:

1. "A Comparative Analysis of Selection Methods Used in Genetic Algorithms"
2. "Genetic Algorithms, Noise, and the Sizing of Populations"
3. "A Gentle Introduction to Genetic Algorithms"
4. "Six Ways to GA Happiness"
5. "Don't Worry, Be Messy"

By the way, genetic algorithms (GAs) are search procedures based on the mechanics of natural selection and natural genetics.* From an engineering perspective they may be used as optimization procedures, and they also have something to say about the theory of design. From a title-design point of view, these five titles run the gamut from fully informative to fully provocative.

The first specimen is about as straightforward and descriptive a title as one could have. Although it is a little long, it does convey a compact version of the talk's contents. Moreover, there is little in the title to offend anyone; a title of this sort is useful when a "Joe Friday" approach is called for ("just the facts, ma'am").

"Genetic Algorithms, Noise, and the Sizing of Populations" is another fairly descriptive title; notice, however, how the use of a triple of topics conveys the breadth of the presentation at the same time that it creates wonder in the reader's mind about how the three topics interrelate. Triples can be overused, but they are an effective device if the juxtaposition is both informative and interest-provoking without being too cute.

The third specimen illustrates how a straightforward title can be made more interest-provoking by the injection of a single word. "An Introduction to Genetic Algorithms" would be a fairly informative, if pedestrian, title. The addition of the single offbeat (and alliterative) word "Gentle" is enough to make the title more inviting. Engineers need to approach the offbeat with caution, however. Your employers may be more comfortable thinking of you as a serious engineer, and it is possible to be too cute. Such matters are tricky, and all I can recommend is that you develop your own good judgment.

The fourth example, with its "six ways," leans even more toward provocation while still being reasonably informative. The actual presentation for which this served as a title is about the use of six elements of practical GA theory to make genetic algorithms work better in applications, and this title does hint at that, creating interest by shrouding the six ways in mystery. (If you're interested in drawing an audience to a presentation of theory, some mystery in one's title is essential.)

The last title, "Don't Worry, Be Messy," goes almost all the way

---

* D. E. Goldberg, *Genetic algorithms in search, optimization, and machine learning.* Reading, Mass.: Addison-Wesley, 1989.

toward provocative at the expense of being informative. The presentation for which I actually used this title combines the material from two separate presentations that were more conventionally named: "Messy Genetic Algorithms: Motivation, Analysis, and First Results" and "Messy Genetic Algorithms: Studies in Mixed Size and Scale." I dared to use such an uninformative title because my audience at the International Conference on Genetic Algorithms was familiar with my work in this area, and I thought that the offbeat title might draw attention to what was essentially a review of material originally presented elsewhere.

As you can see, there is quite a bit of latitude that can be taken in designing an informative, interest-provoking title. As you make more presentations, you will become more proficient picking effective titles. As with all elements of presentation design, if you start from knowledge of your audience, you will not go far wrong.

Foreword: A Word at the Fore. The foreword is an oft-neglected element of a presentation. I use "foreword" (not "forward") to mean an element that sets the stage for a presentation on the larger scheme of things. Specifically, a foreword should contain two elements: background and rhetorical purpose.

Background creates context for a talk. What were the critical events or factors that led to this presentation? In what key ways is this talk necessary? After providing background, it is time to blow the trumpets and give the *rhetorical purpose* of the presentation. Phrases such as "The goal of this talk is. . ." or "The purpose of this presentation is. . ." announce the coming of the rhetorical purpose; the presenter should not be afraid to state what that purpose is. It is important, however, to separate the rhetorical purpose of the presentation from the goals or objectives of the project or the underlying work—they are not the same. For example, a long-term design project may have the project goal of designing a particular gizmo, whereas a project-progress presentation might have the rhetorical purpose of examining specific accomplishments since the last report so that team members depending on the design can adjust or adapt their plans accordingly. Thus the rhetorical purpose has more to do with the expected consequences of the *presentation* than the expected consequences of the *work*.

The foreword can be as simple as a brief statement made while the title slide is on the projector, or it can be a more involved statement accompanied by a more detailed sequence of slides. We will examine the preparation of transparencies shortly.

Overview. Have you ever listened to a speaker who didn't tell you where the talk was going? More often than not, when the speaker finished you didn't know where he or she had been. One of the most important elements of a talk is the *overview*. It should provide a fairly clear route map for the talk: where it starts, where it twists and turns, and where it will end up. Often a single slide will suffice, yet the inclusion of that one slide will do more to help your audience than almost any other. A simple rhetorical device that works well (if it is not overused) is to repeat the overview slide between each major segment of the talk, highlighting the topic you are about to begin. This technique works best in talks where the subtopics are fairly independent. Whether or not you choose to update the route map in this fashion, you should always have an overview slide somewhere early in the presentation; not including it is an invitation to disaster.

Body. The body is, of course, the meat of the presentation, but it is difficult to say much about it in general, other than that the *process* of generating a good presentation body is almost identical to the process of developing good writing. More will be said about the process of presentation development in a moment, but first we need to polish off the elements by examining the summary and conclusions.

Summaries and Conclusions. After you tell 'em what you're going to say and say it, you do need to tell 'em what you said. Specifically, you should do two things. You must summarize the key points of your talk and draw conclusions from the work. I have found that there is much confusion among my students regarding the difference between summaries and sets of conclusions. A *summary* refers to points made during the presentation. *Conclusions* are those consequences for, or changes to, the state of knowledge or the state of the world that are a result of the work presented. In a practical sense, summaries are memory refreshers and conclusions are calls to action (at least calls to changing one's mind). Both are necessary, and both should be presented at the end of the typical presentation.

With the basic elements of a presentation on the table, we are in a position to examine the overall process of presentation preparation in much the same way that we concentrated on the process of writing in Chap. 2.

Preparation Process

We spent most of our time in the writing chapter (Chap. 2) talking about the *process* of writing, and fortunately most of what was said there carries over to the process of presentation preparation. In fact, presentation preparation is so closely tied to the writing process that you'll find it is much easier to prepare a good presentation after you've written something on your intended subject. (It is also easier to write after you have made a presentation on a topic, but we must climb on this merry-go-round somewhere.) Writing the presentation before presenting it does two things. First, it forces you to come to grips with the order of presentation and the transitions between topics. Second, the act of writing programs your tongue for talking. After you've turned a phrase or two on a piece of paper, standing up and giving a talk becomes much easier. Therefore I usually recommend that my students write before they present.

What you write will depend on whether the presentation is *derivative* or *independent*. By a "derivative" presentation, I mean a presentation derived from a piece of writing. Most of my presentations come after I have written fairly extensively on my subject. In such cases, the need for additional writing is limited to the compilation of a list of the topics chosen for inclusion in the presentation and the composition of a paragraph or two on any new topics not previously explored in writing.

By an "independent" presentation, I mean a presentation on a topic that you have not previously written about (or presented). In that case, it is important to go through the mental processes of writing and revision by working up a piece the length of an extended abstract (four to six pages). By doing this as part of the presentation-preparation process, you work through questions of topic selection and ordering fairly fully, and you do enough phrase turning to get some useful tongue programming.

Some may find it sufficient to work up an outline as an alternative to writing an extended abstract. In the writing chapter, I warned that the use of detailed outlines can inhibit creativity and idea exploration, and my warning still holds; however, once the subject matter is fairly fixed in mind, outlining a presentation should not be too risky. Nonetheless, should you find that the ideas are not as fully developed as you thought (as evidenced, for example, by repeated periods of writer's block), return to a fuller exploration of the presentation's flow by invoking the writing process discussed in the previous chapter.

Transparency Design and Preparation

Once you have a feel for the flow of a presentation, you are in a position to prepare the transparency copy and to actually produce the transparencies. We will consider both of these topics after first considering the somewhat peculiar language of transparency copy.

Transparency-Speak. Preparing transparencies may be thought of as an exercise in writing headlines. In other words, the designer of transparencies, like telegraphers of old, must make every word count, sometimes at the expense of complete sentences and other conventions of grammar and usage.

To get in the mood to "speak transparency," I find it useful to grab *The Wall Street Journal* and skim articles by reading only the headlines, both the article and section headlines. If you do this, you will notice the elimination of many adverbs, the suppression of all but the most necessary adjectives, and the use of high-impact nouns and simple verb forms.

We will examine some concrete examples in the next section as we consider the preparation of transparency copy.

Preparing Transparency Copy. With the idea of transparency-speak under our belts, we are ready to tackle the preparation of transparency copy. Usually this involves the writing of a headline and from two to six bulleted topics per slide.

Rather than becoming overly analytical, why don't we do as we did with titles and look at some copy from a presentation entitled "A Gentle Introduction to Genetic Algorithms"?

Let's start with the overview slide copy:

h: Overview
b: Motivation
b: GA basics
b: GAs in search and optimization
b: Advanced operators

   b: GAs in machine learning

Here I've used the shorthand *h:* to denote a headline and *b:* to denote a bullet. Overview slides are usually fairly easy to assemble, as they are simple lists of the main topics of the talk.

Since this is an introductory presentation, most audiences that hear it are unfamiliar with genetic algorithms, and the term itself must be defined. The copy for this defining slide is as follows:

   h: What is a Genetic Algorithm (GA)?

   b: A genetic algorithm is an adaptation procedure based on the mechanics of natural genetics and natural selection.

   b: GAs have 2 key components:

      b: Survival of the fittest

      b: Recombination

Note that only essential information is included and that only a modest amount of information is presented on a single slide.

Further along in the presentation it is important to explain how GAs work. This how-it-works section is preceded by an intermediate overview slide:

   h: GA Basics

   b: Differences—In what ways are GAs different from other search techniques?

   b: Mechanics—How do they work?

   b: Power—Why do they work?

Keeping the listener updated on the progress of a presentation is important, especially in longer presentations. It is also useful as a means to keep yourself on the straight and narrow.

In this same section, another slide gives a laundry list of four ways in which GAs are different from other search techniques. One of those ways is that they are a blind-search technique. The blindness slide copy is presented below:

   h: Blind Search

   b: Canonical search must reject problem specifics.

   b: Treat problem as black box:

   g: [black-box graphic]

The *g:* is used to denote a graphic element (we'll discuss graphics briefly in a moment). Note that because of the inclusion of a graphic element, the text has been kept to a minimum. Too much text on a graphical slide (or vice versa) can be fairly distracting, but you do need some text to help tweak necessary associations that allow the audience to think the right thoughts as they listen and read (and allow you to say the right words as you speak).

Producing the Transparency. The technology of transparency production has changed dramatically in a short time. In the 20 years that I've been making transparencies, I've used pen-stencil sets, pen-lettering devices, pressed lettering, strip-lettering machines, and computers. Of course, computer technology is the most flexible of the lot and is with us to stay. Therefore I recommend that you use some sort of word processor, graphic-arts package, or specialized transparency-making software. The package I've adopted for my own work is SliTEX, which is based on the LATEX and TEX typesetting languages. Most specialized packages should be able to generate decent transparencies if used properly. I like SliTEX because it gives clean layout with a minimum of artistic decision making or fuss.

Again, rather than becoming overly analytical, let's just look at some sample slides created for the presentation discussed earlier. The first is the title slide (Fig. 3.1). As you will notice, it contains the talk title, speaker name, address, and electronic address. Too many speakers begin their talks without mentioning who they are, where they are from, or what they will be talking about. It is a simple matter to ensure that your audience knows you, the title of your talk, and your affiliation.

Scanning this and the other transparencies (Figs. 3.2 through 3.5), you'll see that SliTEX uses large, sans serif fonts. Although serif fonts (such as Times Roman) are commonly used on the printed page, a sans serif font (such as Helvetica) gives an uncluttered look when projected onto a screen. Also, notice that the program sticks to the same font style throughout. This is a rule of good graphic design, and although I am no graphic artist, it seems to me that one of the main side effects of the explosion of new graphics software has been to encourage a lot of junky graphics. The best policy for the nonartist is to keep slide layout and graphics simple. The old army acronym KISS (keep it simple, silly) is appropriate here. You should be warned that engineers as a group tend to be put off by things that

FIGURE 3.1. A title slide should contain the title, the presenter's name, and the presenter's affiliation.

are more glitz than substance. The Apple Computer television commercials, with corporate employees marveling over reports and presentations with extravagant layout, leave the impression that there is no limit to the amount of artwork that should go into the typical business report or presentation. This impression is, to say the least, misleading. The average presentation should be clean and clear; going overboard on graphics will probably have the unintended effect of persuading your boss that you've spent too much time on presenting and not enough time on solving the problem. Of course, if your line of work is computer graphics or visualization, then you will be expected to live up to standards of excellence established in your field. Otherwise, you overdo graphics at your own risk.

Now it's time for the rubber—er, plastic—to meet the road and try some transparency preparation in the next online exercise.

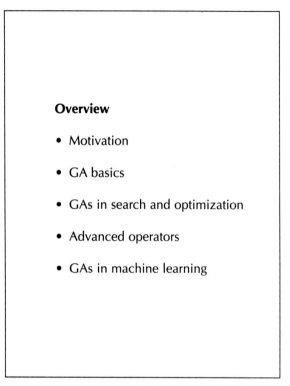

FIGURE 3.2. An overview slide contains a list of topics to come.

Online Exercise 3.2

Using the audience analysis of Exercise 3.1, prepare a seven-transparency presentation on where you'd like your life to be in five years. Choose a title and material appropriate to you and to your real or imagined audience.

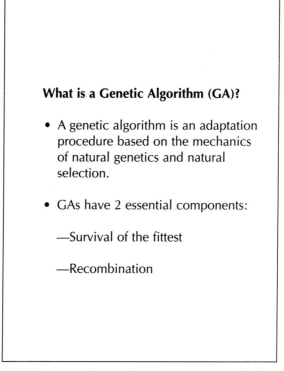

FIGURE 3.3. A defining slide gives a definition of a key term.

In a moment we will focus on how to deliver the presentation you have just prepared. At this point, we need to recognize the similarities and differences between transparency-based and 35-mm slide presentations.

A Note on 35-mm Slides. Until now we focused exclusively on the use of transparencies as visual aids in presenting. Some may wonder why 35-mm slides have not been mentioned. After all, these slides have their advantages: Computer and photographic technology have made them easier to generate; they are useful in large rooms, where 35-mm equipment is better able to project a bigger

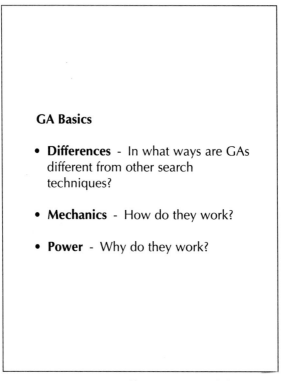

GA Basics

• **Differences** - In what ways are GAs different from other search techniques?

• **Mechanics** - How do they work?

• **Power** - Why do they work?

FIGURE 3.4. An intermediate overview slide gives a list of topics for a section. It is especially important in long presentations to update the route map.

image; and when color is essential or when the presentation relies on photographs, a 35-mm format is cheaper and more convenient.

On the other hand, slides are somewhat disadvantageous as compared with transparencies because of a loss of intimacy. Using 35-mm slides usually forces you to dim the lights below the level required for a transparency presentation. As a result, the audience loses some of its visual contact with you. Moreover, standing behind a podium and punching a projector's remote-control button is a more static activity than being out among your audience, placing transparencies on an overhead projector. In a sense, you trade *doing something in front of* your audience for *showing something to* your

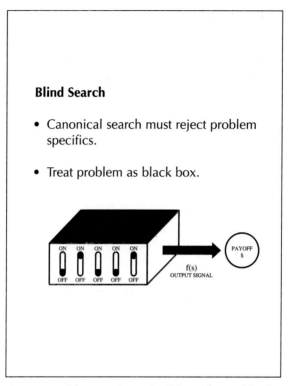

FIGURE 3.5. The sample topic slide with graphic elements shows text and graphics working together. Too much text on a slide with a graphic element can interfere with the overall visual effect.

audience when you go from slides to transparencies, and this translates into a less intimate relationship.

Slides pose a more problematic disadvantage for the occasional presenter: the loss of the cue-card effect of handling the transparencies. When you use 35-mm slides, you must be more familiar with your material and its ordering because you can't preview your next slide before it is shown to the audience. This disadvantage can be overcome by making enlarged copies of the slides for use as notes at the podium, but this puts you back in a situation that is remarkably like speaking from notecards. Also, once you stack the slides in the carousel it is difficult to reorder the presentation. With transparencies you can change your ordering on the fly without your audience noticing. All in all, if you have a choice and you don't have a lot of practice at public speaking, working from transparencies is your best bet. On the other hand, if your company culture—or the material—demands 35-mm-slide presentations, by all means use 35-mm slides; with some additional work and preparation, slide presentations can go as smoothly as transparency-based presentations.

DELIVERY

As I've stated a number of times, one of the primary advantages of presenting over giving a traditional speech is the relative ease of delivery. Presenting lets you work with transparencies, which if properly prepared are actually a full set of notes in disguise; moreover, the setting is less formal and is thus emotionally less demanding on the speaker. Nonetheless, the ability to deliver a presentation effectively is not an inborn trait, and here we consider some key aspects of presentation delivery.

Good delivery requires attention to a number of important details:

1. Voice projection
2. Pace
3. Modulation
4. Eye contact
5. Prop manipulation

The first requirement is that you be heard. Electronic amplification is useful in a large hall, but when using a microphone, you should remember that a normal voice is sufficient to reach the farthest audience member. I once got hoarse while working with the aid of electronic amplification because I forgot I was wearing a mike. In trying to project to the back of the room, I gave myself a sore throat and my audience an earache. Don't make the same mistake; speak at a natural volume when using amplification.

Without electronic amplification, you need to project your voice to the audience members in the last row. The key to projection is to speak deliberately and to enunciate each word separately and clearly, generating a sufficient volume of air from your diaphragm. The amplitude you can achieve approaches that of shouting, but it can be sustained for a longer time. Additional vocal exercises to help you project can be found elsewhere (Wilder, 1986).

It is also important to speak at a modest pace. I learned much about the pace of oral presentation in my six years at the University of Alabama. The Southern oral tradition is widely admired, but until you've settled in to listen to some good ol' tales told in the traditional manner, you can't appreciate how the unhurried pace of the storyteller contributes to the success of the tale. This lesson transfers to presenting. In general, the average presenter is in too much of a rush to get through the material. The audience will understand more when the presenter's pace is in step with the audience's listening speed. Sustaining the vowels in key words seems to help, and liquid consonants (l, m, n, ng, and v) at word endings can be held somewhat longer to create emphasis or dramatic effect.

In spoken English, neither monotones nor singsong voices are much admired. A skilled presenter modulates both the pitch and amplitude of his or her voice to help make the talk easier to listen to; at the same time modulation helps accentuate or emphasize the most important material.

Eye contact is important in building a bridge to your audience, but eye contact can pose difficulties, because the messages received from the audience can cause the presenter to forget what he or she was going to say. Compared to speech making, presenting is fairly forgiving in this regard; you are busy with your props, and the transparencies remind you of the points you are making should you be distracted by eye contact. As you gain confidence in your presenting, you will be freer to look around the room at your audience members; doing so will give you important feedback about their level of understanding and contentment. Eye contact also gives them a further opportunity to judge your feeling and sincerity.

Learning to work with transparencies, projectors, pens, and pointers is not difficult, but the use of such props creates opportunities for the development of annoying habits. To avoid these, it is best to remember a number of dos and don'ts of prop usage:

• Don't block the projected image with your body. A good setup can help: Place the projector relatively high, or project the image onto an elevated screen.

- Don't fidget with transparencies while you speak. Simply place the transparency squarely on the projection surface and leave it there.
- Don't mix notecards with transparencies. One of the nice things about transparencies is that they provide cues to you and information to your audience. Trying to manipulate both notes and transparencies gracefully is next to impossible, so put all important cues on your transparencies and eliminate the need for additional notes.
- Avoid multimedia presentations. Stick with one type of visual aid if possible. If you start with transparencies, stick with transparencies; if you start with 35-mm slides, stay with them. If you must switch back and forth, order the material to minimize such switching.
- To point at an image, point at the transparency with a pen or a pointer. Key words can be highlighted, either in advance or on the fly, by using markers manufactured for transparency application. Should you choose to point at the screen itself, be sure not to block the image.

This may sound like a lot to keep in mind, but good prop usage is mainly a matter of the common sense that comes from trying to visualize your presentation from the standpoint of a typical audience member.

Having reviewed all this, the only way to get better at presenting is to present. This brings us to the next online exercise.

Online Exercise 3.3

Using the previously prepared transparencies, deliver your presentation, taking no more than 10 minutes from start to finish.

Since this is a learning exercise, you should try to get feedback from your audience on all aspects of your presentation: planning, preparation, and delivery. The real world is less likely to provide you with explicit feedback, but over time the reaction of your audiences will give you a sense of whether you are making progress in your presenting.

Summary

In this chapter, we've examined key aspects of making effective business presentations. The key notion is to present—to give a talk supported by visual aids—not speak. Busy businesspeople make a presentation more often than they give a speech, and you should do likewise. We've also considered the importance of audience analysis and subject selection and have enumerated some of the important elements of a typical presentation. This has led to a consideration of the process of presentation development itself, together with a more detailed examination of some of the aspects of transparency preparation, including copy writing and transparency production. Finally, we've considered key aspects of presentation delivery, including the importance of vocal projection, pace, and modulation. While there is a lot to master, by choosing to present (not speechify) you are already well on your way to becoming effective on your feet in front of a business audience.

Offline Exercises

1. Prepare and deliver a brief transparency-based presentation using one of the topics in Chap. 2, Offline Exercise 2.
2. Prepare and deliver a brief transparency-based presentation for the opposing side of the issue you choose in Exercise 1 above.
3. Prepare and deliver a brief autobiographical presentation.
4. Prepare and deliver a brief biographical presentation on an interesting family member or an acquaintance.
5. Prepare and deliver a brief sales presentation for a product or service with which you are familiar.
6. Prepare and deliver a brief sales presentation that pitches your potential as an employee to a potential employer.
7. Prepare and deliver a technically accurate presentation (for a lay audience) on a technical topic of your choosing.
8. Prepare and deliver a technical presentation (for an engineering audience) on a topic of your choosing.
9. Prepare and deliver a brief how-to presentation on some aspect of a favorite sport or hobby.
10. Form a group with five members and choose one of the topics of the previous exercises. In a brief planning session, divide the

topic into subtopics for a group presentation. Prepare and deliver the presentation.

11. Have an audience select a topic of common knowledge. After taking a 10-minute period to organize your thoughts and prepare your transparencies, deliver a brief presentation on the selected topic.

For Further Reading

HOFF, R. (1992). *I can see you naked* (rev. ed.). Kansas City, Mo.: Andrews and McMeel.

WILDER, L. (1986). *Talk your way to success.* New York: Simon and Schuster.

ZITO, A. J. (1963). *Unaccustomed as I am.* Fairfield, N.J.: Economic Press.

DILBERT reprinted by permission of UFS, Inc.

CHAPTER 4

The Human Side of Engineering

Engineering school can be a solitary affair. Long hours doing problem sets with nothing to keep you company but a calculator, a pad of green engineer's paper, and a 0.5-mm mechanical pencil may be good preparation for the technical challenges you will face as an engineer, but the isolation and the almost total self-reliance that result do not help you prepare for the human side of engineering. The meetings, the phone calls, the client contact, and the time spent with co-workers all add up to a level of human interaction that very little in your education has prepared you for.

As I mentioned in the preface, I remember being surprised by the human-relations challenges of my first full-time job, and I did what any red-blooded boy born with a library card in his hand would do: I read. I read about human relations, about sales, about marketing, about organizational behavior, and about leadership. And I made mistakes.

Not little mistakes. Mistakes that lost me friends, mistakes that lost my company sales, mistakes that ultimately sent me packing back to graduate school. And if you're wondering whether someone with such a lousy track record should be writing a chapter on human relations (or more importantly, whether you should be listening to him), so am I. But my reading and my mistakes have led to a somewhat better batting average in this ballpark, and maybe it's better to learn from the .190 hitter who has raised his average to the mid-.200s than from the batter who has always swung a .310 bat. The self-made batter knows something about improvement, while the natural has long forgotten—if he ever knew—how he came to be so good. Of course, I can't guarantee that what I have written in this chapter is true, because life (and the human relations that go along with it) can never be—should never be—a controlled experiment. I can say, however, that whenever I have had the self-control to practice what this chapter preaches I've been successful in my interpersonal dealings.

We start our brief discussion of human relations by forgetting about your favorite person—you—and by trying to look at life through the eyes of others. This is the one axiom of human relations, but a number of theorems and corollaries follow directly from it. We shall look at the role of praise and criticism as well as at the importance of asking questions in dealing with others.

THROUGH THE EYES OF OTHERS

Human behavior is extraordinarily complex, and attempts to simplify the topic risk being naive, ineffective, or both. On the other hand, success as an engineer depends upon a practical ability to figure out the people around us, to work on teams in organizations, and to deal with clients and others outside our organizations. This requires a straightforward approach to understanding and working with people, an approach that does not require a Ph.D. in psychology or an advanced degree in organizational behavior. Yet, with so many different individuals, with their myriad motivations, multitude of life experiences, and variety of temperaments, it seems that any attempt to develop a straightforward approach to human relations would be doomed from the start. What, after all, is common among the individuals we meet in business?

Remarkably, there is one way in which we are all alike, and although it is the source of much conflict between individuals, ironically it is also that which permits us to predict the behavior of others: *We are all self-interested.* Let's face it. You are more interested in yourself than am I interested in you; I am more interested in myself than are you are interested in me. And this holds reasonably true across most pairings of individuals you can name. Of course, there is nothing strange in all this. Biologically, we are organisms programmed to survive, to look out for our own care and feeding above almost all else. As thinking beings, we also devote large proportions of our very large brains to thinking about our higher-level yearnings, wants, and needs. Such natural self-interest sounds counterproductive when it comes to building good group relations, but in a strange way it is the starting point, because if we know that we are self-interested and we know that others are self-interested, we can often predict their behavior.

Thus the starting point of good human relations is *seeing things through the eyes of others.* If we can understand what makes someone tick, we can start to predict what they might do under a particular set of circumstances. And once we have the ability to *model* effectively, as good engineers we know we have the ability to design. To build functioning circuits, we must have a model of how components react and how circuits perform. To build functioning relationships, we must have a model of how individuals behave and interact. Of course, as engineers we use both formal and informal models all the time, but the kind of modeling suggested here is somewhat more inexact and nondeterministic. Comparing our ability to anticipate human behavior to our ability to analyze a circuit, we realize that the human modeling contains more surprises, more randomness and caprice, and much less precision. On the other hand, that our modeling of human behavior is in some ways less reliable than the engineering kind, does not mean it is less useful; perhaps the remarkable thing is that a single straightforward principle results in pretty good ballpark predictions of how people react.

Although the principle of seeing things through the eyes of others is straightforward, its application requires considerable skill. In the remainder of this chapter we consider its use in conflicts, conversations, and persuasive situations. We also consider the important roles of praise, criticism, and apology in our dealings with others.

ANATOMY OF A DISAGREEMENT

Nowhere is the importance of seeing things through the eyes of others more evident than in analyzing the typical disagreement. Sometimes disagreements occur because two (or more) individuals truly have an irreconcilable difference of opinion, but more often than not, one or more of the parties has not taken the time to view the situation from the opposing side.

Consider a case in point. A new engineer in a consulting company was assigned to a project by his boss. The job required that the engineer manage a group of technicians in the mapping of the piping of a chemical plant and the construction of a computer database for the project. The engineer had just completed a graduate degree and was expecting to do work that was more technically challenging than this. His negative attitude toward the assignment came across in his first on-site meeting with the client at the chemical plant when a disagreement arose over the scope of the contract. In trying to get the work to be more technically "interesting," the engineer tried to change the scope of work. This in turn disturbed the client, who complained to the engineer's boss that the work be performed as contracted. Upon returning to headquarters, the engineer got into quite a row with his boss. On the one hand, the boss could not understand how a new employee could be so cavalier with an important client and so bold as to try to change the terms of a contract without authorization. On the other hand, the engineer could not understand how he could be asked to do what he viewed as mere grunt work; this was not the kind of assignment that had been discussed when he interviewed with the company.

I wish I could report that the boss and the engineer resolved their differences and lived happily ever after, but, unfortunately, getting off on the wrong foot set the tone for the engineer's short tenure with this firm. Within six months he left to take another job. It didn't have to happen—if he or his boss had recognized their differences in perception, perhaps the problem could have been worked out. Without a willingness to see the conflict through the other person's eyes, however, there was little chance to reconcile these disparate views.

I should point out that this particular situation arose largely because a person fresh out of school did not consider run-of-the-mill engineering work to be "technical enough." This is a common complaint among engineers and reflects a particular pair of differing perceptions. In school, engineers are exposed to all kinds of fancy technical tools, but in practice the job that needs doing often doesn't draw on that technical tool kit. Thus the engineer is often guilty of not seeing his job through his employer's eyes, eyes that focus on the primacy of the job and the importance of getting it done. Of course, most employers are guilty of not understanding this mismatch between their engineers' expectations and the realities that they face. The engineer's usual sink-or-swim training does little to smooth the road between school and work. This book is an attempt to smooth that road from the new engineer's point of view, but it would be useful as well if employers better understood their engineers' initial orientation and tried to get them to better understand what is required of them.

Beyond these specialized conflicts between new engineers and their employers, it is true that in many arguments, people take firm positions, viewing their side as right and the other side as wrong. There is rarely black or white in human affairs; there are more often shades of gray. Moreover, even when one side is largely right, there is generally no court of appeals to declare a winner and decide how to proceed. In the garden-variety disagreement, if the arguing parties don't work through their difficulty, the knowledge of being in the right can be little comfort in smoothing the ill consequences of the impasse. And in most cases, if there is fault, there is fault enough to go around, so much so that it is useless to try to assess blame. It is better if people try to understand each other's point of view, to separate fact from perception, and to work out a practical way to proceed.

At this point we've done little to solve such misunderstandings. In a moment we'll discuss the handiest habit for encouraging the communication that can bring about greater understanding and

---

Online Exercise 4.1

Consider a recent disagreement in which you were directly involved. Write a paragraph or two analyzing the problem from the other party's point of view. Then consider ways in which the disagreement might have been avoided.

---

fewer disagreements, but before we do that, it might be useful to analyze one of your own recent conflicts from the opposing side.

Conflict is one way in which mismatches in perception manifest themselves. In the next section, we consider how the professional salesperson or persuader can pay close attention to perceptions, thereby minimizing conflict and maximizing agreement in sales and other situations where persuasive skills must be applied.

WE ARE ALL SALESPEOPLE ON THIS BUS

Mention the word "salesperson" to an engineer and you may not get a pleasant reaction. Whether the stereotype we hold comes from the Arthur Miller play *Death of a Salesman* or from our own bad encounters with Willy Loman glad-handers having the gift for gab—and deception—the stereotype does us a disservice, because it prevents us from appreciating and identifying good sales technique.

But who cares? In a book aimed at discussing life skills for engineers, why should we care about good sales technique? Sure, a few among us will earn their pay by pounding the streets of technical sales, but those people receive separate training. Why take time out of our busy agenda to stop and talk about powers of persuasion? The answer to all of these questions is a single word: ideas. As engineers, our primary stock in trade is ideas, oftentimes innovative ideas, ideas that have not been tried, ideas that may encounter stiff resistance from co-workers, bosses, clients, or consumers. These people and others must be *persuaded*—they must be sold—before they are willing to give a new idea a try. Nothing can be quite as frustrating to an engineer as to have a good idea but to be unable to get anyone to take a look at it. This situation is further exacerbated by engineers who often find the logic of their own arguments compelling, so compelling they feel that the world should beat a path to their door. Once again we are confronted by a perceptual problem. The world—our co-workers, bosses, clients, or consumers—often does not see things the way we do—may not think the way we do. If we are to be successful in gaining the acceptance of our ideas—if we are to be successful engineers—we must try to narrow the perceptual divide between us and those we seek to persuade.

Having established that, in a sense, we engineers must all be salespeople at one time or another, we can now ask, What makes a good one? When I worked in the engineering-software business, I hit the streets looking for business, and over the years I had the opportunity to get to know a number of professional salespeople. I found it interesting how far out of line the Willy Loman stereotype was with the behavior of these sales pros. Indeed, the best ones were confident and fairly fearless, but those characteristics were not necessarily the ones that started them on the road to sales success. Usually what separated the stars from the meteorites was their ability to *listen*. Perhaps this flies in the face of conventional wisdom, but a salesperson cannot force you to buy something you don't want. The only real option he or she has is to show you how some product fulfills some physical or psychological need you have. And the only way to find out what that need is is to probe and to listen.

We will consider ways to enhance our listening capability in a moment, but here it is interesting to consider that the stereotypical salesperson—the one who does all the talking—is the antithesis of the effective listener-persuader. The reason we remember the gift-of-gab guys and gals is that they annoyed us so. (Inevitably we walked out before buying anything, or if we did buy, we almost immediately regretted it.) On the other hand, it is easy to forget the good salespeople. They're so smooth we often think that they just wrote up our order, but careful analysis of many such situations reveals an effective listener, matching need with product to facilitate an easier decision.

Let's see if you can extract a human-relations lesson in persuasion from your own recent encounters with salespeople in the next online exercise:

Online Exercise 4.2

Consider two recent experiences with salespeople, one good and one bad. In each case, consider how much they talked versus how much they listened. Compare and contrast how much each salesperson thought in terms of his or her needs versus yours.

Thus far we have considered situations of conflict and situations of persuasion. In both cases, individuals are seeking a change in the status quo, and in both cases a perceptual gap exists. Conflicts arise from a lack of attention to other's perceptions, and sales occur with devoted attention to others' views. The vital question is how does one become more adept at seeing things from another person's perspective? In the next section, we consider the crucial role questions play in this regard.

THE ROLE OF QUESTIONS

The key to human relations is seeing things through the eyes of others, and the key to seeing things through the eyes of others is *asking questions*. Once this is said it is easy enough to understand, but it is surprising that so many people believe that the way to resolve a conflict, make a sale, initiate a friendship, be a good conversationalist, or conduct just about any activity involving others is to tell their side of the story. This approach discounts the interests of the other person, who is as egotistical and self-centered as we and who will be appeased, persuaded, friendly, conversational, or in other ways more positive toward us if given the chance to express his or her views.

The most effective way to draw people out is to ask questions. In this section, we consider the asking of questions in differing circumstances, including conversational, conflict-resolving, and persuasive situations. We will see that different kinds of questions are appropriate in different situations and will identify some of the more important types.

Questions in Conversation

"I'm not a very good conversationalist. I never know what to say."

How many times have you heard someone say something similar to that? No doubt those same people have had good conversations, but it is difficult to stand back from our own human interactions and understand what has transpired. What characterizes good conversation? Usually in good conversation, at least one of the parties asks a question and then listens carefully to the response, following up with more questions that move the conversation along. Thus the prime mover of a conversation is not the talker—that is the easy role. The motive force behind every conversation is the questioner-listener. Of course, the best conversations are those where the questioning and talking roles are exchanged repeatedly.

What types of questions can move a conversation along? It is difficult to generalize, but open-ended questions about something that interests a conversation partner aren't a bad place to start. After all, we know that other people are most interested in themselves.

Questions in Conflict Resolution and Negotiation

Conflict resolution also requires the use of questions but from a more elaborate approach. Whether dealing with an interpersonal problem or an organizationwide conflict, it is important to use various questioning techniques to narrow the perceptual gap that exists between the parties involved. Such a conflict-resolution episode typically begins with the recognition by one or more parties that a problem exists. Once this occurs, one of the parties must observe, "We've got a problem. How do you see it?" or something similar. This fairly open-ended approach invites the other party to share his or her frustrations. When that person finishes, the questioner can briefly summarize what he or she has heard and ask whether the summarized view is valid. If it's not, a more directed question or two can iron out differences and within a few iterations the process should achieve perceptual convergence. At this point, the original questioner might ask whether if it would be all right to share his or her view of the problem. The original talker is now the listener, and after the other view is shared, the listener is asked to summarize what he or she has heard.

After perceptual convergence on this second party's view, a series of questions can then be asked to identify the differences between the two individuals. This series can be followed by a series of more specific questions to see whether there is any room for maneuver or compromise. The progression from open-ended, to

---

Online Exercise 4.3

Take the conflict you discussed in Online Exercise 4.1 and make a list of 10 questions you might have asked the other party to probe his or her position and perception. Make a list of 10 questions he or she might have asked you.

---

confirmatory, to increasingly directed questions moves the parties from conflict toward points of agreement and possible compromise, and closer to the resolution of the conflict. If the conflict cannot be resolved, at least the parties will know that it was not for lack of understanding but rather because of truly irreconcilable differences.

Questions in Sales and Persuasion

Persuasive situations call for all the questioning skill a persuader can muster. We will consider a formal sales cycle as our model situation. In a professional sales situation, questioning usually begins along conversational lines in an attempt to probe the customer's interests, motivations, and needs in a general sense. After identifying needs connected with the product or service, the salesperson may begin a perception-confirming sequence of questions and summaries, and confirmatory-type questions may begin, though there is usually no need for the persuader to share his or her perceptions with the customer.

After establishing a few perceptual outposts, the persuader can narrow the questioning to more specific lines, that is, to what a professional salesperson calls *closing questions*. The bottom line in all persuasion situations is that the business must be asked for and gotten. Books on sales are a better place to read about this well-developed art form, but some of the more salient types of "closes" can be discussed here.

The *conditional close* is a good initial trial balloon, and it goes something like this: "If I can show you that X, Y, and Z occur as a result of using this product, will you buy?" If the person says yes, it is then a matter of persuading him or her that X, Y, and Z will occur.

If the person says no, there is then an opportunity to ask what conditions still obstruct the sale. Along the way, this kind of questioning can lead to the discovery of one or more such *objections*; the uncovering of objections is a call to return to a more open-ended form of questioning to obtain perceptual convergence on the customer's buying blocks, thereby paving the way for their removal.

After removing objections, further closing attempts can be made from the *direct close*, "Would you like to buy this today?" to the somewhat sneaky *assumptive close*, "Would you like it in red or in blue?" In this way, the persuader can work from open-ended, information-gathering questions, to more specific needs-defining questions, to the closing questions that clinch the deal.

Skill in the art of the question can help make us surprisingly effective in our dealings with people. In our increasingly isolated, anonymous society, the art of communication is being lost. It can be regained if we only take time to ask. Another art on the edge of extinction is the art of praising, a matter we take up in the next section.

PRAISE

Beyond the desire to be understood and listened to, each of us loves to be praised. Children adore the praise of their parents. Spouses crave the praise of each other. Workers crave the praise of bosses and co-workers. Despite our ravenous appetite for praise, we are remarkably stingy in handing it out; of course, this represents a remarkable breakdown in our seeing things through the eyes of others.

Why are we so tight with our praise? Do we see it as a kind of currency to be hoarded? Do we view this praising business as some sort of zero-sum game with only so much to go around, so that the praising of others may lead to their success at the expense of our own? Such fears are rarely warranted. Far from being inherently scarce, praise is a fully renewable resource, with many people around us doing praiseworthy things and only ourselves to blame if we don't make the time or effort to notice them.

And it's unfortunate that we *don't* take more time to notice, because praise works a powerful magic on the people it touches. I recall remarking to a frowning, somewhat grumpy woman behind a rental-car counter at Washington National Airport what nice hand-writing she had. A large smile came over her face and we had a nice chat about business and the weather. After this nice chat, and without my asking, she took special care to give me a brand-new car with only 23 miles on it. Understand, I told here that she had good hand-writing not because I wanted a new car or anything else from her. I praised her handwriting because she indeed had lovely handwriting, and what looked as though it might become a stereotypically bad service experience turned into a pleasant human encounter.

Of course, there may be times when you *do* praise with the hope of improvement or change. A young engineer joined a major consulting firm and noticed that the janitorial service in her office was spotty at best. She noticed that the same janitor worked in the department each day, and so one day when the janitor had done a better-than-usual job, she stopped him in the hall and said that she appreciated the extra effort put into the cleaning that day and that she really appreciated it when he took special care in sweeping and dusting her room. The janitor seemed stunned that someone had noticed the extra effort and said something about thinking that no one cared about cleanliness these days. The engineer assured him that she certainly did and that she was glad to have someone working in her part of the building with an old-fashioned attitude toward neatness. Shortly after this conversation the janitor instituted a spot-waxing program in the building, enlisting the help of the other building janitors, and until that janitor was transferred to another building, the engineer never noticed another lapse in cleanliness.

While we should recognize that praise is something we all crave and that it can have a remarkable effect on people, we should guard against that imposter, *flattery*. Flattery resembles praise in that it compliments a person for something, but it lacks *sincerity*. Individuals who have inflated opinions of themselves can be flattered (that is, all of us can be flattered at least some of the time), but in better moments most of us can recognize flattery as the imposter it is. When recognized, flattery can have a worse effect than never having said anything at all.

To distinguish heartfelt praise from flattery, it is helpful to be *truthful* and *specific* when praising. When you say something nice, say exactly what it is you like. For example, in my encounter at the airport, I did not say something vague about the woman's appearance or demeanor; I said I thought she had nice handwriting, and

---

Online Exercise 4.4

Consider a person with whom you have regular contact. Make a list of several things you truly like about that person. On your next meeting, at an appropriate time, offer praise regarding one of the things you most like. Write a brief paragraph reporting what you praised, why, and the individual's reaction.

---

she did. If you say something specific there will be less chance that your comment will be misinterpreted as mere flattery.

CRITICISM

If it is important to recognize those around us for the things they do that we like, it is equally important to avoid criticizing them when they do things we don't like. But when criticism becomes unavoidable, it is important to express it in a way that preserves the recipient's dignity.

That criticism should be minimized is not surprising advice if only we see things through the eyes of others. When was the last time you enjoyed being criticized? I realize, in retrospect, that there were times when I needed to be criticized, but I can't recall a single instance when I was happy or particularly grateful for it at the time. I remember many times when I felt that criticism was overly harsh or disproportionate to the crime, and I can think of a number of people I am less than fond of, largely as a result of their critical nature. My bet is that we all have similar feelings, and the projection of our feelings onto others should be fairly immediate: If we don't like receiving criticism, why should anyone else?

The other thing wrong with criticism is, as a practical matter, it largely doesn't work. Most people have self-defense mechanisms with an enormous capacity to deflect criticism. If armed robbers, rapists, child molesters, and even cold-blooded killers can rationalize their savage, immoral behavior, the average Joe or Jane can certainly deflect accusations of petty wrongdoing. If we are interested in being effective—if we are interested in changing behavior—we must first gain the confidence of the people who have done wrong and help them see why it is to *their* benefit to change their ways.

There are many ways to accomplish this. One is to offer criticism in a spirit of helpfulness. This is a fairly direct approach, and its directness occasionally can lead to resentment; however, words such as "I know you are giving your best effort, and do you think it would be possible if you tried XYZ?" can sometimes temper the blow enough to make a breakthrough. Notice that phrasing the constructive criticism in the form of a question has the effect of tempering the blow even further. Also notice how the use of "and" rather than "but" helps prevent the erection of additional psychological barriers before the sentence is finished.

Another way to temper criticism is to point to your own failings. Sometimes telling a little story about a personal mistake before you ask a person to change his or her behavior is an effective means of offering criticism. It can also be helpful to play down the mistake the person has made. If you make the mistake sound like a big deal, requiring a big effort to rectify, the artificially high hurdle you've erected will make the person resist changing all the more. If you make the error seem easy to correct, you should encounter less resistance to your suggestions.

In addition, it is important not to spend time assessing blame. Some time ago, I had a boss who spent a good portion of his day tracking down mistakes and those who made them. That attitude paralyzed the whole organization, to the point that no one did very much for fear of making a mistake. The proper approach to mistakes is that, once they are uncovered, they be corrected quickly and the individuals try harder to avoid them. Looking to assess blame only makes people more secretive and less cooperative in fixing problems.

It is useful to turn this reasoning on its head at times and use error count as a productivity indicator. The reasoning goes something like this: Given that we try hard to improve on our mistakes, our rate of error tends to remain constant or diminish over time. Therefore the number of errors we make is at least proportional to the amount of work we are doing. We would always rather that errors not occur; but given that they do, and always will, occur, seeing something positive in their occurrence can help us adopt an attitude that contributes to their correction and reduction.

Online Exercise 4.5

Analyze a recent situation in which you were criticized or in which you criticized someone. In two paragraphs describe the situation as it occurred and then describe how it might have been handled otherwise.

It is not only important to avoid assessing blame, but attempts should be made to let a person who has erred save face. There is rarely a need to castigate an individual on a personal level; it is usually the behavior that we don't like, not the person.

ENGINEERING IS SOMETIMES HAVING TO SAY YOU'RE SORRY

While on the subject of mistakes, it seems only fair to consider our own. At the same time we are lightning-quick to point out the errors of others, we can be glacially slow to admit our own.

To recognize the rarity of apologies, we need only ask and answer two questions. When was the last time someone apologized to you? When was the last time you apologized to someone? I don't know your answers to these questions, but for many it has been a long time since they have given or received an apology. What is it that makes us so unable to admit our mistakes, apologize, and move on? Perhaps it is a matter of excessive pride combined with a fear of appearing weak. There is little to do about excessive pride but try to overcome it. On the other hand, apologizing when you have made a mistake, far from projecting weakness, will often be seen as a sign of strength of character and confidence.

One of the reasons we may not want to apologize is that we may feel we are only partially at fault; we think that if we apologize the other person will blame us entirely without examining his or her role at all. In situations such as this, a good approach is the *conditional*, or *explained*, *apology*. In an explained apology you begin by calmly and briefly explaining what irritated you about the other person's behavior, but you go on to say that, regardless of anyone else's

---

Online Exercise 4.6

Consider one recent incident where you felt a person should have apologized to you; then consider one recent incident where you feel you should have apologized to someone else. In each case, consider whether your relationship with each of the individuals has become worse, better, or remained the same. Write a paragraph describing each incident and the change in relationship that occurred thereafter.

---

behavior, you have no excuse for behaving as badly as you did and you apologize. Forced to face having had a role in the problem, many will admit their contribution, and the relationship can proceed with little damage. Others will not see their role, or will refuse to admit it. In these cases, the person making the explained apology can determine whether there is value in using questions to try to get the other person to recognize his or her role in the conflict. In either case, the explained apology can often clear the air sufficiently to allow the business at hand to proceed.

WEAR A LITTLE PASSION

In most situations, it is possible to "view the glass as being half empty or half full." As engineers, our passion for logic tells us that the choice should be a matter of some indifference. As human beings, our passion for passion suggests that we should choose to look on the full portion, because we know that our perceptions of situations can be profoundly affected by our attitudes toward them. Achieving success as an engineer requires persistent pursuit of intermediate and long-term goals over an extended period of time, and such persistence is much easier to sustain if we approach our work—and our lives—with zest and enthusiasm.

People show their passions in different ways, and I am not recommending that we all wear our emotions on our sleeves. I *am* suggesting, however, that some externalization of our positive emotions can have a positive impact on our approach to the challenges of life, and can also help brighten the world of those around us. Organizations that learn to wear a little passion have a positive glow of productivity about them. Of course, it is just as easy, perhaps easier, for organizations and the individuals within them to become gripped by a negative, can't-do attitude. Such working environments are not a pretty sight.

There are a number of practical ways to stay positively oriented; one of the most important is to be doing something you enjoy. We will explore the notion of *engagement* in greater detail in Chap. 6. Another habit is to simply smile and laugh more often. Smiles and laughter are contagious and can have therapeutic value when things aren't going just right.

Another helpful habit is trying to emphasize the good that often eventually comes from initially stressful situations. Most clouds do have a silver lining, and we would all do well to spend more time thinking about the eventual positive consequences of today's mishaps. Moreover, we should make efforts to see the *current* good in bad situations. The doughnut may presently have too large a hole, but there is still an edible portion.

Together these habits can help us get past everyday stresses and obstacles and reach eventual success.

AN ASIDE ON ETHICS

Ethics is a topic that can cause an engineering student's eyes to glaze over, and for good reason. What I remember of the few ethics lectures I heard in engineering school consists of platitudes; wordy, abstract codes; and hypothetical "ethical dilemmas" concocted to stimulate our thinking. I remember those dilemmas as being of two types. There was the "softball" dilemma that always seemed to juxtapose an individual's or a corporation's interests against the greater good of society. We were no dummies; we knew enough to reject that nasty ol' individual or corporation for the sake of the collective. Then there was the "curveball" dilemma: an impossibly complex judgment call involving multiple parties, risk ,money, law, and society. At the time, I wondered whether the softballs weren't a bit too contrived. It often seemed that with a little added complexity the ethical balance could have been tipped in the individual's or corporation's favor. (I now wonder whether there may have been a political agenda underneath the consistent bias against the interests of enterprising individuals and corporations.) I also wondered whether the curveballs, thought-provoking as they might have been, were commonly encountered by the everyday working engineer.

I mention these things in a chapter on human relations, because we are talking here about the primary object of ethics: good human relations. The approach has been unabashedly bottom-up. We have been talking about seeing things through the eyes of others, real human beings of flesh and blood, and trying to act responsibly by doing unto them only as we would have them do unto us. By analogy to economics, we might say that our approach has been a *microethical approach*. It is interesting that ethics is elsewhere taught largely in the abstract—a kind of *macroethics*—encouraging the student not so much to see individuals, but rather to think in terms of larger groups, parties, and interests.

On the one hand, the usual approach to ethics is a logical one. Ethics is an abstracted and codified morality, and so naturally some will choose to approach it initially through abstraction and codes. On the other hand, our ethical failings most often do not result from our lack of knowledge of codes or from an inability to think in the abstract. Do we go astray because of a lack of knowledge? When we lie, cheat, perform unprofessionally, betray a confidence, abuse vested power, or commit some other ethical transgression, is it because we are unaware that it is wrong? Would we have acted differently if we had read the right professional code of ethics? No, our ethical misdeeds come less from a lack of knowledge of what is right than from a failure of execution. We fail to do the right thing because we believe the some other thing will result in personal gain; we are able to do wrong only because we do not consider, or are able to rationalize, the harm our action will cause to others.

Perhaps in discussing ethical issues, it is best to start a little closer to home. If we are studying ethical issues in school, perhaps it would be better to start with discussions of cheating on homework and examinations, the purchasing of term papers, plagiarism, and more garden-variety types of ethical missteps. If we are already in the business world, perhaps it would be best to discuss the padding of expense accounts, the theft of office supplies from an employer, not giving a full day's work for a full day's pay, the presentation of another's idea as your own, and the like. If we confront the everyday matters and develop an ethical approach toward them, the bigger picture should almost take care of itself.

More generally, in approaching ethical questions in engineering design, in manufacturing, in business, or in life, it is better to think of flesh-and-blood human beings. Rather than thinking about anonymous customers, we should think about clients with names, voices, and faces; or we should think about our families and our neighbors—real spouses and real kids. Although no technique is perfect, and humankind will never achieve anything close to an ethical ideal, the concrete approach of seeing ethical matters through the eyes of others is the approach of this chapter and this book.

Having said this, I hasten to add that abstract study of ethics, codes, and case studies can be beneficial once the concrete, interpersonal basis of the ideal is clear. Such matters are beyond the scope of this book, however; several references are provided at the end of the chapter to source materials on engineering ethics and related topics.

Summary

In this chapter, we have examined the importance of good human relations and have considered a number of ways to help us smooth our dealings with other people. The basic principle from which all good human relations flow is seeing things through the eyes of others. Asking questions helps us make the critical shift from our own viewpoint to that of another person. If questions help get us into their minds, praise helps get us into their hearts, helping them feel good about the things they do well.

We have also considered the harm that can come from criticism and have suggested that criticism should be avoided because it is largely ineffective. We have recommended a number of ways to criticize, when criticism *is* necessary. We have considered how apologies should be offered when we discover one of our own mistakes. We have examined some of the reasons why apologizing is so difficult. We have even looked at a way to apologize when another person may also bear some of the responsibility for a conflict. The role of enthusiasm in keeping ourselves and those around us upbeat, positive, and looking forward to the challenges ahead has also been considered.

Finally, we have briefly considered the relationship of engineering ethics to the microethical consideration of human relations. We have concluded that it is best to begin the study of ethics concretely. Abstract ethical study without a concrete foundation can be wasted effort, especially because it is often our ability to think abstractly that permits us to ignore or rationalize the harm of our actions.

In Chap. 8 we will review some of these lessons in the context of organizations. It is natural that the lessons learned in thinking of the interactions between individuals will scale up to organizations, because organizations are themselves composed of individuals. Working in groups poses additional problems that require special attention, but in the next chapter we continue building a repertoire of individual life skills with an examination of the importance of time management and personal organization.

Offline Exercises

1. During the course of a day, make a list of your mistakes, both big and small. Write a short paragraph considering whether the number is larger or smaller than you thought it would be.

2. Keep two lists during the course of a week. On one, record the number of times you are complimented and on the other record the number of times you are criticized. Write two paragraphs comparing and contrasting the quantity, quality, and severity of praise and criticism you experience.

3. During the course of a day, couch all requests for action in the form of questions (for example, "Could you do X?"). During the course of the next day, give all orders as commands (for example, "Do X."). Write a paragraph comparing and contrasting the response to the two approaches.

4. Imagine that you are being interviewed for a job. Make a list of 10 questions your interviewer might ask. Make a list of 10 questions to help clarify, deflect, or redirect an interviewer's questions when they are unclear, unanswerable, or inappropriate.

5. Imagine you are a company representative sent to interview a candidate for a job. Write a paragraph describing specific characteristics of the ideal candidate. Now think of these characteristics from the interviewer's perspective. What ramifications does each have for a potential job candidate's interviewing behavior?

6. At a social occasion, make an effort to hold two types of conversations. In one, make statements and assertions. In the other, ask

lots of questions. Write two paragraphs comparing and contrasting the two types of conversation.

7. Write a brief essay describing the characteristics of an employee your boss would want to have. Discuss the ramifications for your own behavior.

8. An engineer in your group has told you that an engineer on another team has presented an idea of yours as his own. In a short paragraph describe the steps, if any, you would take to handle such a situation.

9. Bill, an engineer in your firm, has written you a "flame," a highly critical E-mail message chastising you for some design work you did a year ago. Should you fire back a return flame to Bill, call him on the phone, see him in person, write a critical memo to his boss, spread rumors questioning his mental state, write a reply via official memorandum, or take some other action? In a brief paragraph describe why you selected your particular course of action. Also, compare and contrast the effectiveness of in-person visits, E-mail, written memos, and the telephone in handling negative situations.

10. Joan, an engineer in your firm, has performed superbly in connection with a project you're working on. Should you ignore her, take credit for her actions, thank her in person, thank her by phone, thank her by letter with copies to appropriate managers, or take some other action? In a brief paragraph describe why you selected your particular course of action. Also, compare and contrast the effectiveness of in-person visits, E-mail, written memos, and the telephone in handling positive situations.

11. Form a group of three and role-play a hypothetical job interview, with two of the group playing interviewers and the third playing the job candidate. Take turns until everyone has played the candidate.

12. Form a group of three and role-play a hypothetical sales call, with one group member acting as the salesperson and the other two acting as prospective buyers at the same target company. Exchange roles until everyone has taken a turn as the salesperson.

13. Form a group of five. Sit in a circle and take turns offering praise to other members of the group. After each statement of praise, the other group members may challenge the remark by saying, "Flattery." If two or more members say, "Flattery," the

praiser gets no points. If one or none says, "Flattery," the praiser receives a point. After five rounds, the praiser with the most points wins.

14. T. Sowell's book, *A Conflict of Visions* (New York: Morrow, 1987), suggests that people come down on opposing sides of many issues because they have radically different *visions* of what is possible in human affairs. He describes the *unconstrained vision* that seeks good outcomes for all and the *constrained vision* that seeks fair *process* for all. For example, a person with an unconstrained vision might look at an uneven distribution of wealth and say that something should be done about it, whereas a person with a constrained vision might ask whether the rules of commerce and employment are fair. As engineers tend to be designers and doers, this might imply that as a group we have a predisposition toward an unconstrained vision. Write a brief essay considering whether this is so or not.

15. M. W. Thring's book, *The Engineer's Conscience* (London: Northgate, 1980), contains six *propositions* that he says are "necessary conditions for the survival of civilization." They are as follows:

   a There is only one humane way of levelling off the world's population and this is to provide a fully adequate standard of living and education to all people.

   b The enormous differences in standard of living and use of resources between groups of people must be essentially eliminated within one generation if we are to eliminate the tensions leading to the Third World War.

   c No pollutant must be emitted to the atmosphere, to water, or to land until it has been proved conclusively that the level of pollution has no long-term harmful cumulative effect on people, animals, or plants.

   d It is a necessary condition for a stable civilization in the next century that the rich countries gradually eliminate their nonproductive activities, such as advertising, weapons manufacture, and fashion and built-in obsolescence, and replace these with genuine attempts to help the poor countries to build up the equipment and knowledge to become full self-supporting at a good standard of living.

   e We have to bring about a fundamental change in the ethos of our society if it is to have any chance of moving into a stable twenty-first century world.

Write an essay debating the merits of any one of these "propositions."

16. A number of the codes of ethics for engineers seek to limit or prevent price competition between engineers. Write an essay discussing alternative explanations as to why such clauses are included in the codes.

For Further Reading

CARNEGIE, D. (1981). *How to win friends and influence people* (rev. ed.). New York: Pocket Books.

FRITZ, R. (1991). *Creating.* New York: Fawcett Columbine.

LUND, P. (1974). *Compelling selling.* New York: AMACOM.

MORRISON, C., AND P. HUGHES, (1982). *Professional engineering practice: Ethical aspects.* Toronto: McGraw-Hill Ryerson Ltd.

RINGER, R. J. (1990). *Million dollar habits.* New York: Fawcett Crest.

SCHAUB, J. H., AND K. PAVLOVIC (eds.) (1983). *Engineering professionalism and ethics.* New York: Wiley.

CHAPTER 5

Getting Organized and Finding Time

Time is one resource we never seem to have enough of; when it runs out, there is no more to be had. Yet, despite its importance and scarcity, it is remarkably easy to waste. To *Homo sapiens*, procrastination is as easy as eating, sleeping, and breathing; even for the engaged worker, getting the most out of the working day is a stiff challenge. To make matters worse, it is remarkably easy to blame our time wastage on external factors and overlook the enemy within. Our clients, our bosses, our families, and our friends are easy scapegoats; although they and others are sources of added work and interruption, the real villain is our own lack of organization, our own lack of discipline, and our own misunderstanding of the pivotal role time plays in our lives. In this chapter, we will examine the many ways people waste time and consider key techniques for gaining control of our schedules and of our lives.

EFFECTIVE WAYS TO WASTE TIME

The wasting of time is an old and venerable activity. Long before the beginning of recorded history, our cave-dwelling ancestors spent time rooting around the cave trying to find the mislaid flintstone so they could start the fire. After many a cave meal many a cave spouse had to nag the procrastinating other spouse to take the carcass outside. With civilization came new technological achievements and advancements in social structure that have helped raise time wasting to its current high art. With alphabets came the opportunity for junk reading, and with movable type came the opportunity for high-volume junk printing. The establishment of regular postal service opened the door for delivery of that junk printing as junk mail. The telephone and the computer have opened new vistas; we now have untold opportunities to send megabytes of useless trash around the world at the speed of light.

Social organization has been no less successful in improving the opportunities to waste time. Kin groups led to tribal organizations to nation-states, to corporations, and, finally, to that most time-waste prone of all organizations, the major research university.

Over these tens of thousands of years the variety of ways to dispose of every spare moment has grown tremendously; nevertheless, it is possible to categorize fairly broadly the ways to waste time:

1. Misplacing things
2. Procrastinating
3. Task switching
4. Never saying no
5. Reading everything
6. Doing everything yourself

In the remainder of this section, we examine these techniques and their effectiveness in some detail.

In a business that generates more than 10 pieces of paper a year, one of the most effective ways to waste time is *misplacing* documents you'll later need. Some individuals are quite systematic in their efforts to misplace important documents. These *pile drivers* have developed an especially effective means of losing any document through the utilization of the *pile document retrieval system* (PDRS). In this system, the user creates several three-foot-high piles of recent and not-so-recent documents. When faced with a need for a particular document or approached with an information-retrieval query ("Do you have memo X?"), the PDRS adherent wheels around to the piles and utters four magic words: "It's in here somewhere." Ten minutes later the PDRSer promises to send a copy when he or she finds it. Sure—and the check is in the mail.

Another proven means to waste time is flat-out *procrastinating*. What the art of procrastination lacks in subtlety it makes up for in unrelenting ability to avoid even the simplest chore. There is room for difference of opinion on this matter, but I feel some of the most creative procrastinators today are those who practice the art in the name of "time management." These people adopt time-management schemes with impossible prioritization plans, using multicolored pens and fancy notebooks with custom-printed forms, only to tell you why the simplest task can't be accomplished in under two fortnights. It is difficult to imagine that procrastination might become any more refined than this, but we shouldn't underestimate the innovative capability of our species.

An equally useful yet somewhat more subtle time-wasting technique is that of *task switching*. Because most jobs require some time overhead to start or restart, task switching maximizes time spent on overhead activities and minimizes time spent on productive ones. When combined with a telephone ringing off the hook and co-workers wanting to talk about the White Sox or the Bears, task switching can achieve near-zero rates of productivity. At the same time, it is the rare task-switching pro who can't get sympathy from co-workers and friends by complaining about all of the balls he or she must juggle.

Another way to make sure you rarely accomplish anything is *never saying no* when you are asked to do something. Even modest-size organizations have a large number of people sitting around with nothing better to do than to generate forms, surveys, report requests, and other trivia to occupy one's time. A perfectly reasonable time-killing strategy is to take all these requests seriously. Fill out that survey on company recreational policy; answer that letter regarding a charitable contribution to Poodles Without Puppies. The skillful practitioner could spend an entire career on information exchanges no more urgent than these.

A close relative of never saying no is *reading everything* that crosses your desk. Important documents, like that 10,000-word article on Zimbabwean frazil ice, should not only receive a close reading from you but may even require your proofing. Not only will this activity dispose of unneeded time, you'll have great fun in recalling third-grade glory days when you were spelling-bee runner-up.

The classic way to prove that you've arrived as a time waster is to try your hand at *broom grabbing*. This maneuver requires that you first hire good people and then do their jobs in addition to your own. This ploy earns bonus points for the successful stylist, because you not only waste your entire day trying to do the work of your subordinates, you completely demoralize and alienate them in the process.

Though my tongue has been firmly in my cheek for much of this section, the satire belies a sympathy for our human condition and our propensity to waste time. We have all been pile drivers and procrastinators, had trouble saying no, read the unworthy, and grabbed the broom. The simple truth is that these time-wasting habits are some of the easiest counterproductive habits to acquire and among the most difficult to shake. Since our natural inclinations work against us, we need a helping hand, a guide to self-discipline; in short, we need a system. In the next section, we'll consider a comprehensive, seven-part system that can help us gain control of our schedules. Although we have no control over the passage of time, we can be in control of its use.

SEVEN KEYS TO TIME MANAGEMENT

With so many opportunities to waste time, it takes a special kind of systematized vigilance to gain control of our schedules and become as productive as possible. In this section, we consider a seven-point plan of attack against the forces of time wasting. The plan involves a two-pronged assault, a pincer movement combining rear action against physical disorganization and a frontal attack against the enemies of productive time use. In this way, we can hope to fight off loss of time and become as productive as we can.

Specifically, the seven parts of our fight against time wastage are these:

1. Find a place for everything and have everything in its place.
2. Work for Mr. To Do.
3. Sam knows: Just do it.
4. A trash can is a person's best friend.
5. Tune your reading.
6. Manage your interruptions before they manage you.
7. Get some help.

Each of these is examined in more detail in what follows.

A Place for Everything

The electronic office is upon us, and far from earlier promises of the elimination of paper, we are awash in a sea of computer-generated reports and are snowed under a veritable avalanche of laser-printed mail and memos. Much of this stuff does not deserve a second glance (and I wish the first were somehow avoidable), but some of it is germane and needs to find a home. The easiest thing to do with all this stuff is to put it in a pile on your desk. As more papers come in they, too, get added to the pile. For a time such a pile-oriented filing system works, because information retrieval from a short stack is not too involved; but as the stack grows, the search time grows as well. It doesn't seem like such a big deal if viewed search by search, but suppose your stack grows to the point where a search for a single document takes an average of three minutes, and also suppose (conservatively) that you need an average of ten documents per day. That means that you spend an average of at least one-half hour per day searching through your piles. Assuming 5 days a week and 50 work weeks a year, this translates into roughly 125 hours, or almost 16 working days a year lost to shuffling through your pile. Almost all of that lost time is avoidable if you build your time-management strategy on the bedrock of a good filing system.

This draws us to an important conclusion: To use time wisely, create and use a filing system. Put in more memorable terms,

> Have a place for everything and have everything in its place.

H. G. Bohn recorded those sage words in 1855, and we could do much worse than to live by them today. There are two reasons—one physical and one psychological—why this is such an important tactic for good time management. I have already alluded to the first reason: Knowing where things are virtually eliminates the pile-driver's pile shuffle, making you that much more efficient almost immediately. The second reason is that by putting things in their place you eliminate the stress of being literally surrounded by pending work. Of course, you must make sure you have a good way of knowing what still needs to be done after work has been filed away—and we will examine one approach to that problem in a moment—but the act of eliminating clutter in your work space can help reduce your worry about the many tasks you need to get done and let you concentrate on the job before you.

I suppose there may be complex theories of how best to create a personal filing system, but the most important things are that you *create one* and that you *use it.* Whatever the system, it should be (1) organized in categories that match your work needs and (2) designed so it is easy to add new files. Over time, it may become necessary to reorganize and recategorize; this will become evident as categories become overstuffed or go underused.

As a concrete example, I have listed the major categories of my own personal filing system:

- Correspondence (by name)
- Student files (by name)
- Course files (by class and year)
- Short courses (by title)
- Projects (by title)
- Proposals; Requests for proposals (by title)
- Papers (by title)
- Paper reprints (by title)
- Personal business (by topic)
- Departmental business (by topic)
- College business (by topic)
- University business (by topic)
- Papers by others (by serial number)

You may wonder why I've shared this in all its gory detail. I remember that when I got my filing cabinet, shortly after taking my job at Stoner Associates, I was curious how other people organized their stuff. How is correspondence filed? How are project and proposal filing done? The system I have presented here is a cross between the things I learned at Stoner and some things I learned from my dissertation advisor. None of it is profound, but sometimes it is easier to design the mundane from other than a blank sheet of paper.

Although the list has been tailored to the needs of a college professor, there are a number of categories of general use. For example, everyone receives correspondence, and a good way to handle it is to have a single category, "Correspondence." This category has individual files for frequent correspondents (by correspondent name) and miscellaneous correspondence files for ranges of the alphabet (A-F, G-M, etc.), where an individual letter is filed by correspondent name in the file folder with the appropriate range of the alphabet.

Student, course, and short-course files are peculiar to my line of work, but project and proposal files are probably necessary in yours as well. It is often useful to distinguish between active and inactive projects and proposals, relegating the inactive kind to deeper storage.

Professors are expected to publish or perish (I prefer the more positive exhortation, "Publish and flourish"), and I keep my original papers in one category and a fresh stack of reprints (ready to go out at a moment's notice) in another. Perhaps in your business it is important to have company literature ready to go out or perhaps copies of past reports or designs. Whatever is important to have available should probably be filed in its own category.

Often there are personal matters that require attention at work (salary review papers, benefits, etc.) as well as departmental and other organizational matters. I keep these in separate categories, and you may find a similar arrangement useful.

Finally, I should mention a word about filing technical papers. It is tempting to keep a file of papers by subject, and this is satisfactory for collections of 100 to 200 papers, but beyond that a more systematic method is necessary. Ben Wylie, my dissertation advisor, files his papers by unique serial number and uses an ingenious system of cross-referenced index cards for retrieving papers by author, title, or subject. I've adopted the same serial-number system, but I use a computer database to cross-reference the file records. I keep a computer-generated author and title listing near my files to facilitate instant retrieval without fooling around with a computer.

However you construct it, a filing system gives you a place to keep things out of your hair and a simple way to retrieve them. Take the occasion of the next online exercise to plan a new or more appropriate file system.

Online Exercise 5.1

Plan a filing system appropriate to your current or future line of work. If you already use a filing system, make a list of its major filing categories. Consider what changes to your system would make it more useful to your current work situation.

Work for Mr. To Do

One of the worries you might have as a born-again time organizer is that if you file something in your spiffy new filing system—if you follow the put everything-in-its-place stricture—you'll be subject to another law of human behavior: "Out of sight, out of mind." Without a systematic means of task tracking, this is a risk. The trick is not to depend on "mind" at all. The trick is to work for Mr. To Do.

We've all used to-do lists at one time or another, but the veteran time manager uses one with no less than religious zeal. Some time-management books suggest elaborate prioritization schemes, multiple lists, fancy calendars, and so on. The tools of our trade will be much simpler: a monthly calendar, a 8 1/2 X 11-inch lined pad of paper, and a diary. The calendar is simply used to record all events, appointments, and fixed-date deadlines. The pad of paper is used to track all activities, both to do and pending, as well as each day's activities. The diary is used to track who you talked to and what you did.

There are many ways to organize your time-tracking pad: Fig. 5.1 shows the simple-minded scheme I use. Rather than elaborately classifying or prioritizing, I list two types of activities: to do and pending. To-do activities are those I need to do within the fairly immediate future. Pending activities are those I would like to do or those that must be done sometime down the road. With this scheme, every morning I review my activities of the previous day and create a list of the current day's activities. As each activity is accomplished, I take great pleasure in crossing it off both the daily and the to-do list; at the beginning of each week, I make a new sheet, updating the to-do and pending entries.

The calendar keeps track of dated activities and deadlines. Whenever an activity with a definite deadline comes in, it goes on the calendar and on the pending list. I check the calendar at the beginning of each week to see which activities are coming due. These are placed on the to-do list for the week, keeping me up to date.

I use the diary to keep track of who I talked to and what I did. To make it simple, I try to make entries in the diary as they happen. Sometimes this results in a somewhat messy diary, but I don't waste time writing and rewriting the same information. I also let my diary double as a technical notebook and do calculations and sketches for new technical ideas there as well. I find that bound notebooks of

MONDAY 7-13		To Do
~~Mike O. Meeting~~		GTD paper
~~MDO Proposal~~		ICGA arrangements
Call ~~TED~~/⟨JH⟩/⟨KD⟩		~~Mike O. meeting~~
		JAPAN Followup
Tuesday		~~MDO Proposal~~
Call ~~JH~~/⟨KD⟩		CCSR update
~~Lunch MWA~~		ICGA AFTERMATH
~~Call Lori C.~~		FGN/NSF REPORT
~~Schedule Graphics for KOA~~		~~PEC Letter for TED~~
⟨Do expenses⟩		Call JH/KD/TED
WED		
Call ~~KD~~		PENDING
~~Do expenses~~		READ VINCENTI
~~Letter for Ted~~		START ECON CLASS
		GE PROMO LIT

FIGURE 5.1. Organization of a simple to-do list can be a straightforward affair.

graph paper with numbered pages work well for both technical and written material. Keeping a written and dated record of your technical ideas is also useful in a patent filing. Pages from a business-technical diary can be notarized, and this is helpful if there is a legal question as to when an idea was developed.

The benefits of this system are several. Compared to many schemes, it requires little or no time overhead: no cards, no fancy priority scheme, just a pad of paper, a calendar, and a diary. Keeping track of your tasks on paper helps clear your mind of the clutter of the many things you have to do. As the use of the filing system unclutters your physical space, task tracking unclutters your mental space. It also gives you a psychological boost every time you cross off a job that's done. Perhaps more importantly, by getting you to face what you do from day to day, it allows you to get a better sense of your productivity potential and to be better able to choose those things that are really important to your work.

With that background, try the following online exercise.

Online Exercise 5.2

If you currently use a to-do list on a regular basis, stop using it for three days. If you do not currently use a to-do list on a daily basis, adopt the scheme just described for a three-day trial period. After the experiment, write a short essay comparing and contrasting your experiences with and without the to-do list. Include a discussion of your perception of productivity, and cite any physical evidence of differences you experienced.

I don't want to bias your thinking, but if I get lazy and leave my list for a day or two, I can hardly wait to get back to it and get control of my working life. One of the easiest ways to keep your list under control in the first place is to get the little stuff done and off the list as soon as possible, and that's the next topic.

Sam Knows: Just Do It

My six years at the University of Alabama were happy and productive, and there I had the privilege of working with a great group of people. One of the lessons I learned came from a memorable mechanics professor and retired Army Reserve colonel, Dr. Sam Gambrell. Sam had the cleanest in-basket of any person I have ever known. Sam's in-basket was so clean that a piece of paper didn't even think about hitting his basket before he had a response completed and shipped out. I must confess that, at first, I thought it was a little silly to be so ruthless about incoming small stuff, but over the years I consciously tried to be a little bit like Sam, and I began to see the wisdom of his ways.

Now I try to handle things as they hit my in-basket. If they're little, ˙ just do them; if they're big, I file them and list them with Mr. To

Do. There is some judgment required here, but I find when I refuse to let regular little stuff pile up, I keep a clearer calendar for the big things that need my attention. I should also confess that I can't quite live up to the example set by Dr. Sam, and from time to time I do get backed up with little things I should have handled quickly. Nonetheless, I know what I should do, and I try to keep clutter out of my basket and keep work flowing out the door.

A Trash Can Is a Person's Best Friend

One constant in business life is that more tasks come across your desk than you can or should do. The easiest way to handle some of them is simply to refuse to do them. The Almighty made junk mail and memos, but he also made pitching arms and wastebaskets, and we should use them. When a potential task first crosses your desk, ask whether you really need to do it; of course, as the new kid on the block, being too fussy can earn you the reputation of being uncooperative, so the new guy or gal needs to tread somewhat warily here. Even so, pay little mind to the junk mail, phone solicitations, and cold calls from salespeople—unless the products they are offering are necessary to what you're doing. Otherwise, let that useful two-letter word—"no"—together with the wastebasket (alias "the round file" or "file 13") unclog your schedule as fast as they can.

Tuning Your Reading

In order to make the "go, no-go" decision just discussed, we often have to read some document, brochure, or other piece of written material to get enough information to know whether something is important or not. In addition, many tasks in the course of the business day have a necessary reading component. On the other hand, just because some amount of reading is necessary does not mean that every business reading task requires the full attention you might put into the reading of a textbook or a novel.

In fact, business reading requires that you have different speeds, that you tune your reading to the task at hand. Unfortunately, years of engineering schoolwork have taught us to read at a methodical and often fairly slow speed, to absorb, for instance, all the material in some fluids or electromagnetics text. In business, this approach is wasteful, and just recognizing that different materials deserve different levels of attention can help.

A simple way to tune your reading is to think of reading at three basic speeds:

1. Skimming
2. Scanning
3. Reading

When you *skim*, your eyes should move from titles, to headings and subheadings, to figures, charts, and tables, perhaps taking in important introductory and summary sentences at the beginnings and endings of appropriate paragraphs. Before reading any document in more detail, use a preliminary skim to derive a road map upon which more detail can be charted during a second pass; this should become a regular habit with everything you read for business.

A *scan* is more comprehensive than a skim; it should cover all elements of a document, not word for word, but as fast as you can while still feeling that you have passed through the whole thing. Speed-reading books teach scanning techniques—for example, the S-curve wiggle down the page and sighting whole word groups instead of single words—and such books and courses can be useful in building your scanning ability. I disagree with some of the speed-reading literature when it makes the overzealous claim that everything can and should be read at scanning speeds. Highly technical material, legal documents, contract specifications, and the like must be read and reread at no faster than near-spoken speeds; and, to be honest, recreational reading is more fun at slower speeds. Nonetheless, scanning is effective for much business material.

I will use the term *reading* to denote your normal textbook reading speed. How fast do you read when it is important for you to remember all of the material that you cover? I think that speed-reading books are somewhat deceptive in this matter by implying that there is little loss in comprehension with faster scanning rates. They "cheat" by defining comprehension as the percentage of correct answers to a superficial multiple-choice quiz given immediately following a reading. Full comprehension cannot be measured by multiple-choice exams; it comes from deeper readings than are possible with a scan. The speed readers do have a point, however, and much of what comes across an engineer's desk deserves little more than the type of scan taught in their books.

Experiment with skimming, scanning, and reading in the following online exercise.

Online Exercise 5.3

Take a five-page document from work, school, or home and give it successively (1) a skim, (2) a scan, and (3) a reading. Time each activity, and after each write down everything you can recall. At the end of all three activities, write a short paragraph comparing the amount learned in the successively more comprehensive readings. Consider whether the incremental knowledge was worth the additional expenditure of time.

In general, it is best to approach a reading task with a skim followed by a scan and, if the information is sufficiently important, one or more readings.

Managing Interruptions

Almost every time-management book that I have read exhorts you to manage interruptions; in a vacuum, this is good advice. The time waste of interruptions is both direct—coming from the expenditure of time on things you hadn't planned—and indirect—coming from the waste of unintended task switching and the associated overhead required to get back to the task at hand. Therefore, all other things being equal, interruptions should be avoided. But all other things are never equal. You don't live in a vacuum; you live in a world of co-workers, clients, bosses, and family members, all of whom have some legitimate claims to a portion of your time. You risk being seen as unfriendly, uncooperative, or worse if you are overly zealous in the protection of your time. If your engineering degree is newly minted, these warnings are especially important, because you are low person on the totem pole, and part of the reason you have been hired is to make life easier for your senior co-workers; being unavailable can be a step toward being unemployed.

On the other hand, disruptions can be managed—albeit carefully. There are a number of ways to hold them at bay. The phone is a primary interrupter, and forwarding your calls to a receptionist or secretary during key work periods can help control phone interruptions, at least for a time. In our electronic age, many people use an answering machine, although some are old-fashioned enough that they prefer that their callers talk to a real person, not a whirling strip of magnetic media. Either way, once messages are taken, you are in control and can decide whether and when to return calls. Of course, your callers may likewise be unavailable when you return their call, thereby starting a nice game of telephone tag. Electronic mail can be useful in this regard, allowing a message to get through to an individual without directly interrupting that person. On the other hand, E-mail opens all kinds of avenues for wasting time, and many electronic messages deserve the electronic equivalent of the round file that receives so much of your other kind of mail.

Unwanted visitors can be partially controlled by closing your door at times when concentration is essential. Here again some caution should be exercised, because an always-closed door can cut you off from your co-workers. (One sign of a healthy organization is that doors are open and people are talking to each other regarding their business.) Another way to control unwanted interruptions, without the negative symbol of the closed door, is to find a hiding place where you can work undisturbed. Much of this book was written in coffee shops and libraries, away from the phone and from my desk. Again, such methods should be used sparingly, lest you gain an unwanted reputation for being unfriendly (or for frequenting coffee shops during business hours).

Within reason, then, interruptions can be at least partially managed, but it is important to monitor the effect of your efforts on those around you. If you control interruptions by disconnecting from your network, you've lost more than you've gained. On the other hand, if you can keep interruptions under control, you're going to accomplish more in less time and be a happier camper, having greater productivity and better control of your schedule.

Getting Help

Another stock piece of advice in the time-management literature is to delegate your work to others. It is true that when you become a manager you must give jobs to others, let them do the work, and avoid grabbing the broom. However, if you're a new engineer, you're not going to have anyone to delegate anything to, and sitting around complaining about that fact isn't going to make you any more productive. In fact, griping about your lack of assistance is one of the worst ways to waste time. Getting the job done is job 1, and to get it done you must first learn the secrets of personal productivity.

Nevertheless, even the new kid may have some opportunities to save time by working through others. Secretaries can be helpful, but in this day and age, with a computer on every desk, it's questionable whether the usual back and forth with a secretary on something like straight typing is the fastest way to go. I find it useful to work with well-trained people who know my handwriting, my filing system, and my preferences, but I find working with pool secretaries to be less productive than doing it myself. Of course, whether this is true for you depends on how fast you type; just make sure that when you get "help" it really does.

Doing rough documents through dictation is another way to use the help of others to make your life easier. At first, using dictation equipment takes some getting used to, but after some practice short letters can be completed on the first trial and longer documents will require only modest corrections. Dictation can be especially useful in initiating the writing process described in Chap. 2. The raw text generated by dictation can be edited, cut and pasted, and interpolated with new text in the generation of a first draft. The principle of creating first and criticizing later still applies, and other than the use of dictation equipment, the process remains the same.

Summary

In this chapter, we've examined ways to waste and to save time. The primary key to personal organization is to "have a place for everything and have everything in its place," which requires the establishment and use of a personal filing system where all your business papers can find their final resting place.

Once you have a place for everything, the next most important key is to work for Mr. To Do. Keeping a simple to-do list of current and pending activities, together with things on today's agenda, will force you to face your time use (or abuse) squarely; the list will act as a kind of higher authority to keep you accountable. Beyond these two key activities, a number of other techniques can be adopted to keep your to-do list more manageable and to prevent unnecessary interruptions from diverting you too much. Some caution is required in adopting time-management techniques that affect others to make sure the activity doesn't weaken your relationship with the people important in your work life; if reasonable care is taken to monitor this, you can save time and be viewed as a team player simultaneously. In the next chapter, we'll consider the importance of effective attitudes toward work and money in trying to find a way to greater happiness—and, perhaps, greater wealth.

Offline Exercises

1. Interview an individual whom you judge to be an effective time manager. Write a brief essay summarizing the key techniques used by that person to help make him or her more effective.
2. Interview an individual whom you judge to be an ineffective time manager. Write a brief essay identifying the key ways he or she wastes time.
3. Write brief paragraphs identifying three ways in which your current organization encourages time misuse. Suggest how things might be changed to remedy the situation.
4. Write a paragraph identifying your stand on the potential conflict between time management and human relations alluded to in this chapter. Is there a conflict? Which is more important? It might be helpful to consider specific incidents, hypothetical or empirical.
5. Create a personal filing system and use it for one month. Write several paragraphs describing your experience.
6. For one week practice what Sam Gambrell preaches, doing little things immediately and filing others for further consideration. Write a brief paragraph describing your experiences.

For Further Reading

FRANK, S. D. (1990). *Remember everything you read: The Evelyn Wood 7-day speed reading and learning program.* New York: Avon Books.

MAYER, J. J. (1990). *If you haven't got the time to do it right, when will you find the time to do it over?* New York: Simon and Schuster.

REAM, M. L. (1977). *The Merrill Ream 10-lesson speed reading course.* Mission, Kan.: Sheed Andrews and McMeel.

DILBERT reprinted by permission of UFS, Inc.

CHAPTER 6

Money, Work, and You

Money. Just the word has the power to cause us pause, to make us think. Many of us think that our prayers would be answered and that our lives would be at peace if only we had a little bit more.

Unfortunately, it is rarely that simple; our feelings about money (or our lack of it) are often complex and sometimes mask much deeper concerns about our careers, our lives, ourselves. In this chapter, we will investigate some of these concerns by examining three of the more common roads to wealth. Two of these are fairly well known, whereas one is something of a surprise—but the truly interesting thing is what they share in common. We'll find that the three roads to wealth are paved with a good match between what one likes to do and what one actually does. Along the way, we'll see how wealth is also facilitated by a match between earning and spending styles. And we'll dispel three myths about how wealth is obtained— myths that, if believed and pursued, can harm one's prospects for personal happiness *and* wealth. Finally, we'll consider the importance of cultivating the habit of thrift.

THE ROADS TO WEALTH

How do wealthy people get that way? Well, some get their money the old-fashioned way: They inherit it. Others win the lottery or are otherwise graced with good fortune, but we are less interested in these cases than those where the fortunate have had more of a hand in amassing their fortunes. Those working stiffs who find their way to financial security find many ways to get there. A study by Blotnick (1980) followed 1057 people from 1960 to 1980.* In that time, 83 of the 1057 people (7.9 percent) amassed personal fortunes in excess of 1 million dollars. In his study, Blotnick looked for similarities in spending habits, saving habits, earning modes, and even sexual habits to see whether there was something about this group that made them special. What was common to the 83 people that helped them along the road to wealth? Did they share a particular line of work, such as sales or real estate? Were they particularly good at investing? What was their secret?

Among the things Blotnick learned was just what the members of this group were *not*. They were *not* members of any one profession. Stereotypical notions of the wealthy salesperson, real-estate broker, and so forth were found to be untrue, though to be sure, there were members of these professions among the wealthy. The group was also *not* noted for its investing skills; even when individuals showed talent at investing, it was not a primary factor in their attaining wealth in the first place.

The study did show that the group earned its wealth in one of three ways, two fairly conventional and one fairly surprising:

1. Starting or building a business
2. Rising through the ranks of an established organization
3. Accidental investing

The first two routes need little explanation and are not surprising. The third does need some explanation and is almost a complete surprise—a surprise that gives a clue to the common thread running through all three of these paths to financial security.

The study showed that a fairly large number of individuals became wealthy as a result of *accidental investing*. These individuals

---

* We will cite this work throughout this chapter, referring to it as "the study" or "the Blotnick study."

purchased an asset—often real estate—in connection with their primary work activity, and after a number of years (typically a large number) the asset became valuable, and the individual sold out at a substantial profit. For example, the study tells of a woman who as a hairdresser ran her own shop, eventually buying the rundown building in which it was located for $75,000. Over a long period of time, the neighborhood became more fashionable, and she sold out for $900,000. One might say she was just plain lucky, but—and here is the kicker, the interesting thing about "accidental" investing— look at what she was doing between buying the building and selling it. She *wasn't* checking the ads every week to see whether property values had gone up, like a normal investor or speculator might do; nor was she trying to make a quick buck. She *was* simply cutting hair, an activity she enjoyed very much; and because she was so engaged with her chosen line of work, she sat on her piece of property long enough for it to become valuable. Had she watched the pot, it certainly would never have been given a chance to boil; in a sense her accidental investing was born of *unconscious patience* that grew out of satisfaction with her work.

This is a remarkable finding, and it would have been enough if that were all that Blotnick discovered, but after finding that accidental investing grew out of patience, which grew out of love of work, he noticed that people who traveled the two conventional roads to wealth—starting or building a company and rising through the ranks—were also involved with and enjoyed their work deeply. We will call this deep level of involvement and enjoyment of one's work *engagement*. It turns out to be the single most important factor in career happiness and the increased probability of long-term financial success.

To demonstrate the effects of engagement in the two more conventional roads to wealth, the study cites a number of anecdotes, including a story about two brothers who, starting with a $10,000 investment, founded a wire and cable company in 1957. The firm grew to the point that in 1980 it was doing more than a billion dollars worth of business. In thinking about what had made their business grow, did these guys credit their desire to be wealthy? No, they credited their engagement, citing the early days when they would work Saturdays, driving around looking for and buying wire and materials. In short, they enjoyed their business. To them it was totally engaging.

The study also cites the example of a warehouse manager who ran a clean, efficient, and organized shop. One day, an executive from a much larger company, a supplier, visited this warehouse and was surprised to see an operation much better than his own, an operation better than any other he'd seen. The executive hired that warehouse manager at more than twice his previous salary, and the warehouseman went on to work his magic at the larger company. Did he do it for the money? No, in fact the manager had operated his warehouse that way long before there was any hope of substantial financial reward. He did it because he found his work engaging.

Now wire-cable companies and warehouses may not be your cup of tea, but the message is general and applies to each and every one of us. Regardless of whether you work for others or work for yourself, and regardless of what type of business you are in, from high-tech, to low-tech, to no-tech, getting engaged with your work greatly improves your chances for financial success. Of course, not everyone who is happy on the job becomes wealthy, but therein lies the hidden beauty of putting happiness first. If you do work that is deeply satisfying, so what if you don't make a million? You're doing something you enjoy and making an income you can live on. Aren't these substantial rewards in themselves? That your chances for greater financial security are enhanced is something of a bonus. In short, engagement matters. In a moment we will examine why engagement matters so much, but first we'll examine whether there is life beyond work.

IS THERE LIFE AFTER WORK?

Having sounded the clarion call for engagement in one's work, I should hasten to say that this does not preclude a family life or other outside interests. All work and no play does both a stale engineer and a lousy family person make. Of course, there will be times in your life when work will be quite demanding and outside interests will suffer, but such times will be balanced by others when family duty calls and work has to take a back seat. Knowing how to manage this shifting balance is difficult, and there is little that can be said in general about these matters, except that we should try to avoid using one of life's facets (such as work or family) as an excuse for neglecting another.

For example, workaholics don't necessarily put in long days because they are engaged with their work. Sometimes they are running away from problems at home. Likewise, there are people who devote their waking hours to clubs, investments, or other outside interests, showing little involvement with their work; oftentimes such people are running away from an unsuitable work situation. It is important not to deceive oneself about one's problems, whatever or wherever they are. Sometimes simply confronting a problem directly can improve the situation.

Even when we are able to live a fairly balanced work-home life, the demands of juggling many balls require that we manage time well, particularly at work; the suggestions of the previous chapter can be helpful here. Many workaholics—far from being time-efficient—base their long workdays on a foundation of poor use of time. Knowing that you're going to be around later takes the pressure off doing something now. On the other hand, a balanced existence that forces you to divide the day among various facets of life can actually help you to better allocate your time in each facet.

One way to squeeze more out of your day is to rearrange your sleep schedule. Ben Franklin rose at 5 A.M.; that same habit can be useful in more modern times. By rising at five and going into work shortly thereafter you can (1) have uninterrupted time to write and think first thing in the morning, when you are freshest, (2) stay until a normal quitting time and squeeze more out of the working day, and (3) spend a full evening with your family or friends without worrying about work. Of course, there is some evidence of a physical basis for both larklike and owlish behavior, and decisions about scheduling are very personal matters. Nonetheless, sometimes our sleeping habits are dictated by external factors or conventions rather than conscious thought; some thought about your own circumstances, followed by a conscious decision, might generate a schedule more suitable to your needs.

WHY ENGAGEMENT MATTERS

Having seen that the need for engagement must be balanced with the need to have a home life, we return to recognize that being involved with one's work is the basic factor in becoming wealthy and successful. But why does it matter so much? We could examine complex psychological reasons, but keeping to our hard-nosed engineer's approach, we identify two simple factors. Engagement is an important determinant of success because

1. Time flies when you're having fun.
2. Details aren't annoying when they are part of an enjoyable bigger picture.

We've all had the experience of time "flying" when we were engaged. When was the last time it happened to you, and what were you doing? For me, my most recent time-flying experience is this current one: As I write these lines, the hours are passing, but I hardly notice them go. Maybe times flies for you when you work with your hands, program a computer, read, write, do a tough problem, or participate in a favorite sport or hobby. It really doesn't matter which activities cause time to fly for you; it matters that you know which ones they are and try to find ways to make them a part of your work.

Of course, the knowledge that time does fly when you're having fun is no secret, but why is it a determinant of success? We've already seen how accidental investing breeds a kind of unconscious patience that allows an asset to become accidentally valuable; but time flying is also a key to the other two roads to wealth. Whether you build a business or climb the corporate ladder you must develop your personal and organizational competence before success will knock on your door—and that development takes time. Considering how impatient human beings are, it is that much easier to stick to one's knitting and pay the dues that need paying if the passage of time doesn't seem so tedious.

Another thing about being engaged is that it takes the devil out of the details. All jobs worth doing have myriad details that need to be done—and guess what? People who are engaged with their work don't mind doing them, because they are involved with getting the total job done and therefore have a good attitude toward what otherwise could be tedious labor. By way of contrast, the disengaged worker finds the details annoying and wishes the work could be delegated away. Although this lesson applies at all levels of society, as individuals we can only exercise reliable control over ourselves (and even that is often in doubt). Therefore the best bet is for each of us to strive to find a work situation that offers the opportunity to make a contribution to an effort that engages us. Situations that match our inclinations aren't altogether easy to find, but they needn't be that difficult either. In a moment, we'll discuss some practical ways of finding a good fit, but right now we have to question whether ours is an inherently engaging profession.

IS ENGINEERING AN ENGAGING PROFESSION?

The other day, when I was lecturing about engagement to my GE 291 senior seminar class, I asked for a show of hands by those people who were thinking of getting out of engineering. To my surprise, roughly half the class members raised their hands. Of course, these are tough economic times, and perhaps the high proportion of potential switchers was simply due to the tight job market. It is also possible, however, that the large number indicates the many students who get to the end of the engineering-school line and still don't understand what it is they like—if anything—about engineering.

One reason this can happen is that an engineering education does not necessarily reflect how engineers actually spend their time in the workplace, and so it is possible to get to the end of such an education and still not understand the joys of engineering. This book started from the premise that an engineering education ignores much of the communication and people skills required to get the job done; but even if we restrict our attention to the technical side of an engineer's education, we note a mismatch of sizable proportions. Real engineering is results-oriented and is concerned with prescription—what should be done; it is less concerned with the process used to do it. Engineering education, on the other hand, is almost entirely process-oriented and is almost totally absorbed in description (modeling, analysis, applied science, or whatever you may want to call it); many engineering students who ultimately make very inventive and innovative designers find that they are only so-so at the stuff considered important in school. Likewise, many of the students who do well at analysis find the ambiguity of the actual art of engineering less than comforting.

Beyond this mismatch between technical education and the real world, it is also true that as engineers we are such busy beavers that we rarely step back from building our dams to contemplate which logs we enjoy chewing.

Engineering is a multifaceted profession and can be engaging because it is

1. A creative profession
2. An intellectually stimulating and challenging profession
3. A real-world profession
4. A constructive profession
5. A people profession
6. A maverick's profession
7. A global profession
8. An entrepreneurial profession
9. An optimistic profession

First and foremost, ours is a creative profession. As engineers we create that which has never before existed, through a combination of imagination, ingenuity, and perseverance. We therefore have many opportunities to become engaged in the creative processes of idea generation and problem solving. This stands in stark contrast to those professions—the medical profession is just one example—that train their practitioners largely to become proficient in extant technology and technique.

Engineering is also an intellectually stimulating and challenging profession. Being a good engineer requires much knowledge and know-how, but no armchair intellectuals need apply. Ours is a profession that requires streetwise application of mind to means, where the touchstone of success is whether the job gets done.

This leads us to recognize that our profession is firmly rooted in the real world. This has a number of benefits. It forces us to face up to the limitations in our modeling, and it forces us to confront difficult variables that defy analysis—variables such as time, money, consumer preferences, the impact of government, and the impact of technology on society.

Moreover, engineering is an inherently constructive profession, attempting to make a better world through change. Contrasting this to the legal profession, we note that law oftentimes adds cost to many transactions without adding direct value to the products and processes it touches. Engineers often find great pleasure in being able to touch or see the results of their labors, taking great pride in their contribution to a completed product or project.

Ours is a people profession, as engineers often work in teams. As marketing, manufacturing, and engineering considerations are integrated into the design process, engineers increasingly find themselves working on teams with many different types of individuals across a company. Of course, we've devoted a good bit of space in this book to emphasizing the habits necessary for good interpersonal relations, whatever the circumstances; but the engineer who is skilled in his or her dealings with others will also find many opportunities for engagement therein.

At the same time that our profession requires team effort, it can also call for outstanding individual effort. Many of the most creative and advanced engineering projects have required a champion to almost single-handedly overcome obstacles and single-mindedly bring an idea to fruition. Thus ours is a profession that finds a place for the engaged maverick at the same time it embraces the team and its players.

Our world has become a very small place. Jet travel allows us to become physically present almost anywhere in the world in less than a day. Satellites, fiber optics, and advanced computing allow us to become virtually present almost anywhere in the world in nanoseconds. Such changes are making engineering a more global profession, where products are designed and built across borders, even across oceans. This situation creates opportunities for the engineer who is willing to learn other languages, customs, and cultures.

Some of the same technological influences that make engineering a more global profession are opening up new entrepreneurial opportunities. Engineering has had a long tradition of private practice and private enterprise, but the tumult that is modern electronics and computing has opened new vistas for the engineer-entrepreneur. As communications technology makes close ties at a distance a reality, more and more engineering functions will be farmed out to independent design shops at remote locations. At the same time, the tools of our trade have dropped in price; the small shop need be at little or no competitive disadvantage to the in-house engineering operation of a Fortune 500 firm. Moreover, machine-tool and materials-handling technology is driving manufacturing toward point of sale; we need only look at the photo-processing business for hints of the future of manufacturing. As we move in such directions, it should be clear that a company's competitive advantage will lie more in its software—in its designs—and less in its manufacturing and distribution capability. Although the present has been kind to the entrepreneur-engineer, the future holds many engaging opportunities for those with the enterprising spirit.

Finally, it almost goes without saying that ours is an optimistic profession. Our impulse as builders is reinforced by the knowledge that we have improved what was once a very hard life and the hope that our continued efforts can make things even better. Sometimes we have paid insufficient attention to the unintended consequences of our acts, but the genie of innovation prefers freedom to the confinement of the bottle, and once free he has largely served his masters well.

Thus there are many different ways to plug into the engineering profession. Yet finding the things that engage us as individuals is still a difficult task. In the next section, we consider ways to better match who we are with what we do.

MATCHING YOUR PERSONAL IMPEDANCE

The lesson of engagement is both liberating and vexing. It is liberating to know that the road to wealth is paved with happy effort, but it is vexing to try to find our way to work that makes us happy. How can we better match our inclinations to the work we take up? Additionally, how do we match an occupation's earning style to our spending style? Both of these questions are important in finding what we'll call our *personal impedance match*. Just as electrical engineers know that circuits work best when impedance values of two components match, in the real world we plug into life better if our interests and inclinations are consonant with the demands of our work. We examine the elements of this impedance match and learn to go about finding it.

Activities and Skills Inventory

We've noted that time flies when you're having fun; but what activities keep you jolly, and what skills or tasks make up those activities? An entire aptitude-testing industry has arisen to answer such questions, but our approach to uncovering our inclinations will be much simpler. Let's write our way to an answer with the following exercise.

Online Exercise 6.1

Write an essay that explores three specific instances when time flew and three instances when time crept. Do not limit yourself to work activities. Include activities around the house and leisure activities as well. Explore the various skills and activities involved in each instance, and try to assess which contributed to time's flying or creeping.

The "time flying and creeping" criterion is the key to this exercise, because it gets you to focus, not on whether you were consciously happy, but on whether you were engaged enough that you did not notice the passage of time. As a secondary criterion you can evaluate how annoying details seemed. These two guidelines should help you perform some meaningful introspection.

Earning-Spending Impedance

Finding activities that tickle your fancy is important, but another important question of fit compares how an occupation pays wages with how you like to spend them. At the risk of oversimplifying, we'll say that wages are paid in two ways: in *clumps* or in a *trickle*. For example, commission salespeople get paid in clumps: Make a big sale, get a big paycheck. Freelance writers, artists, and consultants are similarly paid. The person starting a business has periods of clumplike payment, although after companies get underway, they themselves may be divided into clumpish and tricklish types of companies. On the other hand, salaried men and women see their wages in trickles; every week, every two weeks, or every month they receive a modest but regular sum of money. These are two very different ways to receive money, and neither is intrinsically better than the other. The trick, as the Blotnick study pointed out, is to match the way you receive money with the way you like to spend it.

Blotnick classified spending styles in three broad categories:

1. High rollers

2. Low rollers
3. No rollers

Although the term "high roller" usually has a negative connotation, the study used the term simply to mean anyone who likes to buy many things and has little or no difficulty in making such purchasing decisions. "No roller" was used to describe those who do not like to buy things and sometimes agonize over even fairly modest purchases. "Low roller" was used for those who make thoughtful, considered purchasing decisions without excessive agony.

As I've intimated, your spending style is best matched to an appropriate earning style. That is, high rollers will be comfortable with the clumps and no and low rollers with the trickle. Problems can arise when high rollers choose to work for a trickle, or when low and no rollers choose a job that pays in clumps. One very telling example in the study was that of an engineer, a low roller, who forced himself to go into sales because "that is where the money is." The bottom line was that he was always anxious about the unpredictability of his income. Of course, the message of this example is not that engineers should not become salespeople. It is that engineers—like everyone else—should try to find occupations that allow a good match between their earning and spending styles; radical mismatches can result in economic difficulty, unhappiness, or both. With this in mind we are ready to explore spending styles in Online Exercise 6.2.

Online Exercise 6.2

Write a short essay where you consider three recent purchases: one minor, one moderate, and one major. Consider the decision-making process that went into each purchase. Was it impulsive, considered, or agonized? How long did the decision take? Did you comparison-shop? Did you buy top-of-the-line or bargain-basement? After considering the three purchases in some detail, classify yourself as a high, low, or no roller.

Work Analysis

Having taken stock of our skills and our spending style, it is time to consider the match between our inclinations and our chosen line of work in Online Exercise 6.3.

Online Exercise 6.3

In an earlier exercise you considered your goals over the next five years. In the last two exercises you considered the activities that make time fly and your spending style. In a short essay, consider whether your career plans help create a personal impedance match to your inclinations and spending style. If so, which elements of your goals should be emphasized to enhance the impedance match? If not, what career options might offer a better impedance match and why?

These are difficult questions, and I should warn you that your preferred skills and activities—and your spending style—can change over time. Old dogs *can* learn new tricks; but even so, performing an analysis such as this, now and at intervals throughout your career, can help you see if you're currently on a plausible track. Should you find that there is a mismatch, the exercise may help you find your way to a better match. Even if you find that your personal impedance matches that of your chosen work, there are a number of pitfalls to avoid.

THREE DEAD ENDS

There are three myths about wealth that can deceive people into wasting much time, effort, and money:

1. Compete and you will grow rich.
2. Socialize and you will grow rich.
3. Invest and you will grow rich.

In the remainder of this section we will briefly examine each of these.

Dead End 1: Compete and You Will Grow Rich

We are a competitive lot, and one of our more persistent myths is that the fiercest competitor will claw his or her way to the top. The Blotnick study showed that combatively competitive behavior can work initially, if one is pursuing the climb-the-ladder way to wealth. (It is largely ineffective on the other two roads to wealth.) Examples were given where individuals clawed past their colleagues by overtly aggressive, sometimes unethical, behavior. This strategy worked because the promotion decisions rested with individuals too far from the front lines to witness the often shameful behavior of the promoted individuals. As these people were continually bumped up, however, they reached a level close enough to the locus of power that decision makers could witness their antisocial behavior firsthand, and at this point the competitive climbers were blocked from further rise. Thus the strategy is partially successful in the short run, but in the long term, overtly competitive behavior is self-defeating because organizations function best when people work together, with common respect and courtesy toward each other and mutual support of the group's mission and goals.

I should be quick to acknowledge that the kind of competition I am talking about here is the overt, aggressive kind. Hard work and hustle are much admired in business (and elsewhere), but crossing the line from working hard to attacking a colleague's ideas, calling someone else's work your own, or generally gaining at the expense of a co-worker represents defections from corporate cooperation that cannot long be tolerated.

Dead End 2: Socialize and You Will Grow Rich

Another interesting conclusion of the Blotnick study was that socializing one's way to success is largely a dead end. Blotnick traced a number of social climbers and professional party goers and found that their efforts, too, were partially successful, sometimes leading to interviews or tryouts. Yet, the inordinate amount of time these people put into appearances—the time invested in achieving form over substance—hurt them when the time came to perform. Their interviews often led to rejection; and even when they were hired, their lack of competence led to disappointment or worse.

Having rejected overt socializing as a road to success, we should not go overboard and reject the importance of what others think about us and the role of a good interpersonal network. Scheele (1979) makes this point well when she classifies people into those who are *sustainers* and those who are *achievers*. Sustainers spend most of their lives doing a good job and the rest of it waiting for recognition and rewards that don't come. Achievers spend most of their lives doing a good job and spend the rest of it letting the world know what it is they've done. It goes a little bit against an engineer's logical grain to have to spend much time promoting his or her efforts; we have a sustainer's tendency to expect competence—especially technical competence—to be rewarded on its own. Unfortunately, it doesn't always work that way, and it is useful to have a mentor—someone who counsels you and watches out for you. It is also a good idea to keep your bosses and your colleagues informed of your accomplishments. In the wider world, it is useful to become active in professional societies, through membership, publication of papers, or other activity. In these ways, when you do something of interest, it is more likely to be noted, which can lead to recognition and possible career advancement.

Dead End 3: *Invest and You Will Grow Rich*

For some reason that I've never understood, there's a lot of pressure to invest one's money in stocks, bonds, puts, calls, futures, bellies, straddles, or whatever the financial instrument *du jour* happens to be. Perhaps this pressure comes from a self-interested financial community, or perhaps it is natural in a capitalist economy—but, whatever the reason, the pressure is there. It is so ingrained that sometimes we are driven to the mistaken conclusion that investing is the primary way to wealth and that work is the way one sustains the habit of investment. For many it is a costly habit indeed. Wealth usually comes (if it does) during the engagement phase, when an individual is engrossed in his or her primary activity and succeeds by climbing the ladder, building a business, or accidentally investing. In the subsequent phase, the investing phase, now-wealthy people have money to invest and do. The problem is that the rest of us observe the latter phase—nobody paid much attention to these people until they became wealthy—and mistakenly think that their current behavior is responsible for their great wealth. This is a widespread myth, and one that most of us would do well to discount. This is not to say that we should not take care of our money, and in a moment we'll discuss several concrete aspects of thrift and saving, but the primary suggestion of this chapter is to find work you enjoy and get good at it.

It is interesting, if not surprising, that these dead-end myths persist, because each activity has some payoff some of the time. In the parlance of optimization, these behaviors represent *false peaks*, or *false optima*, on the landscape of success; like most false peaks, if your search is too local, you will remain stuck at the false optimum. Whether the strategy endorsed here—finding a match between your inclinations and your work—is a globally optimal strategy is an open question. But there's little question that it's a pretty darn good one. Moreover, the message is simple and positive. Instead of chasing dollars, one should seek happiness and perhaps the dollars will flow. Either way, you'll have lived a better life.

THE HABIT OF THRIFT

Having argued against investing as the primary mode of generating wealth, I should backpedal somewhat and argue for developing the habit of thrift. Before taking a flyer on Peruvian grape futures or other exotic speculations, it makes sense to accumulate an amount of money that equals or exceeds one-half of your annual salary as a way to protect yourself against unexpected financial problems. Moreover, having cash in the bank gives you a cushion should you choose to switch jobs, go back to school, or otherwise do something that interrupts your income stream. In thinking back to my own decision to return to graduate school after having worked for a number of years, I know that it was made easier because I had saved a good portion of my earnings.

Once you've saved some money, it's fairly natural to ask whether you are getting a reasonable return. One of the best returns on investment comes not from an investment but from smart shopping. To understand this, consider that if you can get 25 percent off the price of a $100 pair of shoes you need, that savings is equal to the amount of interest you would have received had you invested the $100 in an instrument bearing 25 percent per annum. There are few (legal) investments that regularly pay 25 percent per year, but there are many opportunities to save 20 to 50 percent on items that we need every day. Of course, one of the tricks to this is not letting your thrifty consumerism lead to unnecessary or unplanned purchases; nevertheless, thrift in purchasing certainly can help put money in the bank.

In order to understand what to do with your money once it is in the bank, it is helpful to develop mental models of financial investments and how they work. To do so requires reading, and I've already recommended *The Wall Street Journal* as a good newspaper on its own merits, but if you become interested in things financial it is *the* newspaper. *Forbes* is a good choice among business magazines, because the writing is lively, the news is current, and the tone is irreverent. The classic book in economics self-education is Hazlitt's gem, *Economics in One Lesson*. To get a good overview of various types of investments, Harry Browne's *Why the Best-Laid Investment Plans Usually Go Wrong* is a good place to start. His categorization of financial assets as cash, gold, stocks, and bonds is a useful simplification, and his writing is as straightforward as it is sensible. You might disagree with his rigid balanced-portfolio notion, but his suggestions for separating your permanent, or investing, portfolio from your speculative portfolio are wise, as are his warnings about the difficulty of beating the market through agile timing.

The basic message of this chapter is that investing your money is secondary to investing in yourself and developing core competence at work that you find interesting; nonetheless, there are basic decisions to make in the allocation of your savings. The information provided here is simply a pointer to some useful reading that will help you make decisions suitable to your circumstances.

Summary

In this chapter, we've examined the three roads to wealth: (1) starting and building a business, (2) climbing the corporate ladder, and (3) accidental investing. We've found that, regardless of the chosen path, the common key to success is *engagement* with one's work.

With that in mind, we've seen that the easiest way to a happier, perhaps wealthier, life is to find work that matches our interests and provides payment in a manner that matches our spending style. When we achieve such a *personal impedance match* along these dimensions, we stand a much better chance of finding both happiness and financial success.

Even with an impedance match, the obstacles to happiness and wealth are many. In some ways we have seen the enemy and he is us. Specifically, we have warned against trying to compete, socialize, or invest to success. Although each of these strategies offers some short-term gains, they ultimately consume time, effort, and money that might be better spent on happy, productive work consistent with who we are and who we can become.

Offline Exercises

1. Interview a successful person. Ask that person questions about his or her early years and about how he or she became successful. Ask questions about that person's present work life. Write an essay discussing the keys to his or her success. Compare and contrast levels of engagement and enthusiasm early and late in the person's career.
2. Form a group of two. Each of you should write a brief statement about time-flying activities and interests. Exchange lists, then write a brief essay on specific career options that seem to match the other person's list. Read and discuss both sets of recommendations.
3. Form a group of two. Each of you should make a list of major purchases made over the last two years. Exchange lists and write a brief essay classifying the other person as a high, low, or no roller. Discuss the essays.
4. Form a group of three or more and choose one individual whose career choices will be examined during a brainstorming session. (See Chap. 7.) Try to find alternative career paths or emphases that will help the individual to grow. This exercise can be repeated for each group member.
5. Form a group of five and have each person make a list of major purchases made over the last two years. Make copies of all five lists and circulate them among the group members. Rank the lists from highest to lowest rollers and discuss attitudes toward spending money.

For Further Reading

BLOTNICK, S. (1980). *How to get rich your own way*. New York: Playboy Paperbacks.

BOLLES, R. N. (1979). *What color is your parachute?* Berkeley, Calif.: Ten Speed Press.

BROWNE, H. (1987). *Why the best-laid investment plans usually go wrong.* New York: Morrow.

FRANKLIN, B. (1961). *Benjamin Franklin: The autobiography and other writings.* New York: New American Library.

HAZLITT, H. (1979). *Economics in one lesson.* New York: Arlington House.

SCHEELE, A. (1979). *Skills for success.* New York: Ballantine Books.

TOBIAS, A. (1979). *The only investment guide you'll ever need.* New York: Bantam Books.

CHAPTER 7

Groups, Innovation, and Structured Brainstorming

Meetings are a necessary evil in business life, but they need not be as evil as they often are. Although there are many different types of meetings—regular staff meetings, one-on-one meetings, huddles, sales meetings, and client meetings, to name a few—the most important, and the most nettlesome, is the problem-solving or brainstorming meeting. Given the long-standing existence of effective procedures and protocols for productive brainstorming sessions, the number of organizations that have *not* adopted any of these methods is surprising. In this chapter, we will probe the anatomy of a typical problem-solving session. Be warned: It is not a pretty sight. We will then examine the reasons for the ineffectiveness of the typical meeting and outline a six-step brainstorming procedure that can unlock the innovative potential of any group. The adoption of this single technique can boost substantially the creativity, quality, and quantity of the solutions generated by even the stodgiest of organizations.

A DAY IN THE LIFE OF A TYPICAL MEETING

To set the stage, imagine that you've just learned of a difficult problem facing your organization. Your group leader calls a meeting to discuss the difficulty, and at the appointed hour the troops gather. The high-priced talent sitting around the table knows the importance of this moment. The meeting begins. The group leader briefly outlines the difficulty and throws the meeting open for ideas. One group member begins, first raising several aspects of the difficulty that the group leader omitted from his problem description, then proposing a specific solution. A second group member challenges the effectiveness of the first member's solution on the basis of cost and also raises several other unreported aspects of the problem. She finishes her statement with her own proposed solution. A third group member finds the first two solutions inadequate on the basis of implementation time and suggests a third solution. And on it goes.

The meeting continues in this manner, with group members proposing and rejecting each other's solutions, until finally the group leader, realizing that the meeting is going nowhere, takes the bull by the horns. As a good leader, the best he can do at this point is to piece together a solution from the best suggestions he has heard. If he is an autocrat, he simply announces *the* solution, a solution that the careful listener heard in his opening and subsequent remarks. Either way, the group members leave with something of an empty feeling, knowing that, yet again, the group's creativity and innovative potential have gone largely untapped.

WHAT'S WRONG?

The scenario described above is all too familiar in business, but the reasons for the failure of most unstructured brainstorming are fairly easy to pinpoint:

1. Inadequate discussion of the various facets of the problem
2. Premature criticism of partial solutions
3. Mixed discussion of solutions and criteria or objectives
4. Inhibition of full exploration of ideas by a leader-follower dynamic

In the remainder of this section we'll examine why each of these factors inhibits group brainstorming productivity.

The first difficulty in unstructured problem solving is that the problem does not usually receive an adequate initial airing. In our imaginary scenario, the meeting leader started off with his view of the problem, but other people around the table obviously had additional information and different viewpoints that could have been helpful in understanding the problem more fully. Since unstructured meetings allow any kind of discussion at any time, as soon as the meeting is opened up to the group, natural human impatience almost guarantees that solutions will be proposed before the problem has received adequate discussion. This is detrimental to a group's ability to come to consensus because there is no shared vision of the range and complexity of the problem.

The usual unstructured meeting has a serial idea-criticize-idea-criticize rhythm to it, which is particularly harmful to a group's problem-solving productivity. Finding a solution to a problem usually requires the recombination of a number of notions from a number of sources to arrive at something that works. If notions are disposed of before they've had a chance either to generate or inspire other solutions or refinements, or to be recombined with other partial solutions, the end result will be less satisfactory than it otherwise might have been (that is, had more partial solutions survived to be considered in the final analysis).

The unstructured nature of the typical meeting also leads to a mixed discussion of partial solutions along with criteria for judging those solutions. Premature rejection of partial solutions is often accomplished by raising a particular criterion by which the proposed solution is judged more or less inadequate. Of course, this disregards the other eight criteria that the solution may fully satisfy and points out the difficulty of viewing solutions in isolation with particular criteria. There are almost always trade-offs to make in choosing solutions to tough problems, and it is better to postpone the consideration of all solutions in the context of all criteria than to use individual criteria as bullets to shoot down each new idea that dares to raise its innovative little head.

The last-but-not-least difficulty in many meetings is the leader-follower dynamic. In many organizations, leaders are accustomed to making decisions without much input from those they lead; typically, a manager asks for input only to unveil the "correct" answer at the end of the "brainstorming" session. Thus many seasoned veterans come to such sessions with the idea of listening for hints about the Politburo's chosen solution rather than listening and contributing to a genuinely creative experience.

Formal managers are necessary in all organizations, and these managers have the right and duty at times to exercise their decision-making authority. When they call a brainstorming session, however, they have a moral obligation to their people to listen to their ideas and fiduciary responsibility to corporate shareholders (or organizational backers) to try to reach the best solution possible. This requires some restructuring of the way meetings are conducted, if only to overcome the tendency of leaders to lead and followers to follow.

The four difficulties of unstructured brainstorming—inadequate discussion of the problem, premature criticism of alternatives, mixed discussion of solutions and criteria, and the workings of the leader-follower dynamic—remind us of the difficulties faced by the individual writer (Chap. 2) as he or she tries to get thoughts on paper. There writer's block can be attributed to trying to create and criticize simultaneously. In a group problem-solving session the same conflict between creating and criticizing arises, but the size of the group complicates and intensifies the destructive reaction between the creative juices and the critical venom. Many meetings turn into survival-of-the-loudest (or longest-winded) sessions or, worse, a contest where only the biggest boss's ideas get considered. In the next section, we'll examine a structured approach to brainstorming that separates the creative and critical thinking throughout the problem-solving process, thereby permitting group productivity and creativity to flourish.

STRUCTURED BRAINSTORMING

After sitting through dozens of meetings and witnessing hundreds of good ideas being shot down in the usual fashion, one begins to wonder whether there might be a better way to solve problems in a group. I know I was ripe for my first encounter with brainstorming techniques when I took a software sales training course in Indianapolis in the late 1970s. The brainstorming protocol taught was based on an adaptation of Alex Osborn's (1963) original protocol; it's been used widely at General Electric and other innovative firms. I have since seen several variations on the same theme, but I will present here the variation I remember. Let's take a look at the props, personnel, and rules required for structured brainstorming.

There are a few physical props required for structured brainstorming:

1. Large flip charts
2. Adhesive tape
3. A large felt-tip marker

The flip charts are used to record the proceedings of the brainstorming session; adhesive tape (masking tape works best) is used to secure the flip-chart sheets to the walls in the meeting room; and the felt-tip marker is used to record the proceedings. Sometimes people suggest that a blackboard be used, but this is inferior to flip charts, because most rooms do not have enough space to record a full meeting and blackboards are not a permanent record. If a session is going well, dozens of flip charts can be filled in a matter of minutes, and it is very easy to get a meeting record of 25 to 50 sheets. It is also unacceptable to forgo a common public recording of the proceedings; it is important that individuals know and agree to what is being recorded, and it is important for everyone to have access to the "shared memory" of the full meeting record at any time.

Brainstorming works best in groups of three or more, and during the session all members of the group must have equal status, whether they belong to management or to the rank and file. Initially, one person volunteers or is designated to be the scribe. He or she stands at the flip chart and records the proceedings as accurately as possible. The scribe is not a meeting leader and is authorized only to record, not filter, information. He or she is allowed to ask questions to clarify a point so it can be accurately recorded and may call attention to procedural matters, but while a person holds the marker as scribe he or she is prohibited from making a creative or critical contribution to the session. This rule is very important: It prevents the scribe from slipping into a "leader" role by prohibiting him or her from taking too active a part. To be fair, however, a scribe wishing to offer a contribution to the session can become a participant by handing the marker to another group member, who then becomes the new scribe. Scribeship, in this way, can and should be shared by many group members.

With proper props and a scribe in place, the structured brainstorming session can begin. In its normal course it follows six steps in the following order:

1. Discuss the mess.
2. Define the problem.
3. Generate solution alternatives.
4. Generate and select solution criteria.
5. Rate alternatives according to the criteria.
6. Select a proposed solution or set of solutions.

In the remainder of this section, we'll discuss what is meant by each of these steps and why they are so ordered.

Discussing the Mess

Sometimes problems are simple enough that short briefings by a group leader are enough to understand the difficulty. More often than not, however, real-world problems are fairly complex, requiring the input of many group members to flesh out their full extent. During this initial phase of brainstorming, members discuss the mess; that is, they bring up background information, historical information, the present situation, other solutions that have been tried, and any other information that may help the group understand the difficulty. During this phase it is important to require that there be no debate or argument. Group members may independently present their different views of the world and all views are recorded by the scribe. There should be no attempt to organize the material in any way, and members should be encouraged to associate freely. Contributors should be brief and to the point; long speeches and war stories should be discouraged. Additionally, some effort should be made to avoid being prescriptive at this point because there is as yet no problem to solve. There will be ample time for generation of alternatives later in the process; it is more productive during this phase to concentrate on symptoms of difficulty and hypotheses regarding root causes.

After discussing the mess for a time, it is common for members to begin to sense that the group's wheels are spinning. Issues are repeated or packaged in new wording, but little new information is coming out. Such repetition is often a signal that the mess has been sufficiently discussed. At this point, the scribe or any other group member sensing a slowdown can ask whether the group is ready to define the problem. It may be useful at this point to go back over the session history, to see if other ideas are generated associatively. Once the group is ready to move on, it is time to define the problem.

Defining the Problem

During the discussion of the mess, many issues, both germane and peripheral, are invoked. During the second phase, it is time to focus on which issues should be tackled during the remainder of the session. In terms of the writing model of Chap. 2, discussing the mess is analogous to directed creation, and defining the problem is analogous to revision. One difference between the writing model and a problem-solving meeting is that the meeting is a group activity, and it is important to come to a succinct statement of the problem that the group can agree to. One way to avoid unnecessary conflict is to tend toward inclusiveness in the problem statement. At this stage, as long as the problem definition is fairly well on target, a somewhat larger definition that encompasses the views of the whole group is better than one that arbitrarily excludes some issues important to a significant minority. Thinking ahead to later stages, it may be possible to find solutions that cover those special concerns without much extra effort or cost; even if it isn't possible, the extra concerns can always be discounted at the later stage of formulating a solution or set of solutions.

During the problem-definition stage, discussion is permitted but care should be taken to avoid bickering and needless debate. Once a suitable, succinct problem statement has been created, it is time for generation of alternatives to begin.

Generating Alternatives

Generating alternatives is the most exciting part of structured brainstorming. The rules are simple: No holds are barred, all ideas are welcome, and no criticism of any idea is permitted. Not one word. This rule is the moral equivalent of "not crossing out" in freewriting and directed writing. The scribe simply writes down alternatives as they are generated, posting each new sheet of alternatives on the walls around the room. Group members are allowed—they are encouraged—to bounce ideas off one another to create hybrids or embellishments; ideas should flow freely and associatively from one to another. Again, there should be no effort to make ideas come out in any particular order. Human thought is a messy process, and we should let it be so. There will be plenty of time for the harsh light of reality to shine on silly or infeasible alternatives. In the meantime, every outlandish, wacky idea that gets mentioned increases the chances that some creative, innovative, and perhaps more practical idea might pop into someone's mind.

If the problem is sufficiently difficult and more than one session is required, it is often useful to break alternatives generation into multiple sessions. If group members have a chance to "sleep on it," they often return to the table refreshed with alternatives that would not have occurred to them during a single session.

Generating and Selecting Criteria

Solutions are only good and bad in relation to criteria that they satisfy or don't, and in this step of the brainstorming session a set of criteria for judging the alternative solutions is generated and selected. During discussion of the mess, many criteria typically get mentioned, so the first part of this phase is to collect candidate criteria by reviewing the "mess." Additional criteria can and should be added to the list. As with the other creative portions of the procedure, this first pass at generating criteria should avoid criticism and be inclusive. Thereafter a second pass should be made to ensure that each included criterion is essential to project success. If controversy arises, it is still best to err on the side of inclusion. In the final debate, spurious criteria will usually be discounted.

Once the list is culled, a decision must be made on how to score each criterion. This can be as simple as a qualitative judgment of relative effectiveness (+), ineffectiveness (-), or indifference (±), or a simple subjective score (say, 1 to 10). Some criteria lend themselves to a more quantitative evaluation of a relevant statistic, such as expected profit, volume, sales, and so on. Once the choices are made, each alternative solution can then be rated according to the list of criteria.

Rating Alternatives According to the Criteria

Once the alternatives are generated and the criteria are chosen, alternatives can be ranked according to each criterion. The easiest way to go about this is to make a matrix with alternatives listed down one side and criteria listed across the top. In the usual alternatives-generation session, some of the solutions will be basic configurations, and others will be features or refinements that can be added to (or taken away from) one or more base configurations. For example, in solving the problem, "obtaining personal transportation to and from work," base configurations might be

1. Buy a vehicle.
2. Lease a vehicle.
3. Take the bus.
4. Ride a bicycle.
5. Walk.

Refinements that might naturally arise during the generation of alternatives could include a listing of specifications of the cars that might be purchased or leased, the type of purchase or lease plan, arrangements for maintenance, and so forth. When listing each feature in the evaluation matrix it is convenient to group features with their base configurations so they can be considered for inclusion or exclusion independently.

Selecting a Solution or Set of Solutions

After the matrix is filled out, it is time to do some deciding, or at least some culling of the list. There are formal decision-making procedures for using multicriteria ratings such as these, but often the process of going through the brainstorming exercise and filling out the evaluation matrix will sufficiently focus the group's attention on the best solution or solutions. If this happens, great! If not, it is likely that two or more subgroups feel that there are significantly different solutions that are best for the organization. In these cases, the best thing to do is *not* to seek compromise. Subgroups espousing different solutions should hammer out separate proposals, and the final decision should be made in a manner consistent with normal organizational decision-making procedures.

Sticker Voting: A Quick-and-Dirty Shortcut

Some decisions deserve the full brainstorming treatment as

described above. For others, either the costs of having a group sit around and go through the entire procedure are too great or action is required fairly quickly and cannot wait for the full procedure to run its course. In these cases, there is a useful abbreviated scheme that can quickly determine whether there is consensus on the outline of a solution. In this shortcut, called *sticker voting*, the brainstorming process begins with the first four steps: (1) discuss the mess, (2) define the problem, (3) generate alternative solutions, and (4) generate and select solution criteria. (If time is really pressing, even the fourth step can be dropped.) Thereafter members of the group are each given a set of colored stickers and are asked to vote by placing their stickers directly on the flip-chart sheets next to the elements of a solution they favor. A somewhat chaotic scene usually follows, with members placing their stickers, horse-trading votes, and bumping into each other as they make their decisions. After the dust settles, large clusters of stickers identify the most-favored elements of a solution, and the number of sticker votes may be recorded and passed along as the group's recommendation. The shortcut relies fairly heavily on the group members' intuition regarding the connection between solution elements and criteria; by not going through the formal process of considering solution elements against each of the criteria, it is possible that group members will miss important trade-offs in making their evaluations. Nonetheless, in cases where a quick or inexpensive recommendation is necessary, abbreviated brainstorming with sticker voting brings the benefit of a full discussion of the mess and alternatives without the protracted evaluation of solutions against criteria.

PUTTING STRUCTURED BRAINSTORMING TO WORK

The process we have examined is fairly straightforward, and if the rules of engagement are followed closely, the result can hardly help but be an improvement over the usual serial idea-criticize-idea-criticize approach adopted in most unstructured meetings. To get some experience using the technique, try the next online exercise.

The beauty of structured brainstorming is in its ruthless separation of the creative and critical components of the process, as well as its prevention of the leader-follower group dynamic. Practiced reg-

Online Exercise 7.1

Apply structured brainstorming to the problem of finding employment as an engineer in economic hard times.

ularly, it can help boost the quantity and quality of the solutions created by any group with which you are associated.

Summary

In this chapter we have examined a method of structured brainstorming that can replace the usual setting up and shooting down of ideas that occur in many business meetings. Specifically, a six-step brainstorming procedure separates creative from critical activities, permitting ideas to inspire and be recombined with other ideas, later allowing the ideas to be selected according to the multiple criteria appropriate to the problem at hand. By employing a nonparticipating scribe who creates a shared public record of the transactions, the procedure largely avoids the leader-follower dynamic, and ideas are considered on their merit, not on the basis of who suggested them. The result is better meetings, better solutions, and better group acceptance of the solution or solutions selected.

Offline Exercises

1. Form a group of three or more members, and apply the structured brainstorming procedure (or the abbreviated scheme with sticker voting) to "solve" the following problems:
   a. Schools in the United States graduate students with little or no mathematical-scientific-technological literacy as compared with schools in other industrialized countries. Design a curriculum that addresses this problem.
   b. According to recent surveys, cheating among college students is on the increase. What steps can be taken to address this problem?
   c. Although engineering influence remains strong in some high-tech sectors of the economy, many technologically intense corporations are dominated by individuals with a background in finance or law to the exclusion of individuals with product or manufacturing know-

how. Is this a problem? If so, what steps can the engineering community take to reverse this situation?

   d. The engineering curriculum seems increasingly disconnected from what engineers do in the real world. Design a curriculum that addresses this problem.

2. Select one of the topics in Exercise 1 and try the structured brainstorming process as an individual. After going through the steps of the process, write a report detailing your solution and your experience with the structured approach.

3. Form three or more teams consisting of three or more individuals each. Select one problem from the list in Exercise 1 and have each team solve the same problem using the structured brainstorming protocol. After the solutions are complete, convene a meeting of all teams and compare and contrast the selected solutions.

4. Select a problem that was recently solved in your organization by traditional, unstructured means. Perform structured brainstorming on the same problem, individually or in a group. As much as possible, ignore the previous solution. Compare and contrast the solutions derived by structured and unstructured means.

For Further Reading

EMMERLING, H. (1992). *It only takes one.* New York: Simon and Schuster.

MICHALKO, M. (1991). *Thinkertoys.* Berkeley, Calif.: Ten Speed Press.

VON OECH, R. (1992). *Creative whack pack* [deck of cards]. Stamford, Conn.: U.S. Games Systems.

VON OECH, R. (1990). *A whack on the side of the head: How you can be more creative.* Stamford, Conn.: U.S. Games Systems.

OSBORN, A. F. (1963). *Applied imagination.* New York: Scribners.

DILBERT reprinted by permission of UFS, Inc.

CHAPTER 8

Organizations and Leadership

For much of this book we have focused on the individual skills necessary to function as an effective engineer; to operate at a higher level—to design, start, or run an effective organization—much of what is needed is simply the application of good principles of individual conduct at the organizational level. However, since organizational design and leadership are such highly leveraged activities—since they affect so many people—it is important to get them right. Moreover, it is important to understand whether the tightly knit group setting of the organization imposes any new constraints on

121

what we've learned so far. This causes us to examine why mutual commitment and cooperation are so important, yet so hard to achieve.

Beyond the application of human-relations principles at the organizational level lies the important question of *strategic vision*. Can leaders see through the day-to-day hustle and bustle of the business enterprise and perceive what currently makes their companies successful? Do they sense what challenges loom to threaten a company's continued existence? Do they recognize opportunities as they knock, and can they open the doors to continued growth and prosperity? In discussing questions of vision we are necessarily on less firm territory than when we discussed more tangible topics. We are surveying important ground, because it is vision that ultimately determines whether a company continues to prosper in a competitive environment; it is vision that must constantly be renewed in the face of the comfort of old habits.

In what follows, we first revisit the others'-eyes principle of human relations in the organizational context. Dealing with people one-on-one is difficult enough, but generalizing from such dealings to uncover principles of organization seems almost impossible, if only because the range of human behavior is so wide. This leads us to consider a unifying model of human motivation and two theories of management that result from the model. We ask why there are organizations and this in turn leads us to the important concept of cooperative enterprise. Thereafter we'll consider a number of key ideas in organizational design, as well as a number of habits of leadership that seem to hold up in practice. We'll conclude with a few words about vision: how to find it, how to articulate it, and how to share it with others.

THE RANGE OF HUMAN BEHAVIOR

In Chap. 4 we considered relations between pairs of human beings and concluded that the key to good human relations is the others'-eyes principle. One-on-one, this seems like a sensible strategy. Now we must generalize to a collection of individuals, even though it is difficult to see how we can draw general lessons from what we know about a particular individual. We will start by asking if there are bounds on behavior that help us delimit what people might do in a particular situation.

Looking on the brighter side of human nature we know that, among other things, people can be

- Gentle
- Contemplative
- Insightful
- Punctual
- Neat
- Polite
- Trustworthy
- Creative
- Energetic
- Consistent

Looking on the darker side, we also know that people can be

- Brutal
- Oblivious
- Uninspired
- Tardy
- Sloppy
- Rude
- Untrustworthy
- Dull
- Lazy
- Inconsistent

Apparently, there are few bounds on human behavior, and, to make matters worse, we can find the best and the worst of these behaviors exhibited by a single individual.

Looking at behavior alone is of little help in our attempt to model groups of human behavior, but this should come as no surprise, for when we consider the others'-eyes principle from a distance, we recognize that what we are really talking about is *motivation* not *behavior*. This leads us to consider a simple yet useful model of human motivation in the next section.

A UNIFYING MODEL: MASLOW'S HIERARCHY

There is a straightforward model of human motivation—Maslow's (1970) hierarchy of needs—that is a favorite of management theorists and practitioners. The model is popular because it is relatively simple to convey and fairly easy to apply in practical situations.

Maslow's model begins with the premise that human beings are needs-driven animals and goes on to say that human needs are organized more or less hierarchically, from basic to more complex (as depicted in Table 8.1). Thus, as a person fulfills basic bodily needs (such as food and oxygen), he or she begins to focus on higher needs (such as safety). As these are increasingly satisfied, needs for social interaction become more important, and so on down the list (which is up the hierarchy).

Models are useful to engineers for their real-world predictive power, which this one seems to possess. Applied to the bewildering array of behaviors surveyed in the last section, it makes it possible, perhaps, to understand what motivates a behavior that we might otherwise be tempted to label "good" or "bad." For example, when an employee is evasive and defensive in the face of repeated questioning by a negative, distrustful boss, it is easy to understand the employee's fear of losing a job and to see how basic needs such as food and security are at stake in the conflict. Likewise, it becomes easier to understand the extra effort exerted by an employee who has been repeatedly praised by a positive manager if we understand the ego gratification such praise provides.

Of course, organizations and their leaders interact strongly with the needs hierarchy of all their individuals, and it is interesting that as we move from more basic concerns (body, safety) to higher needs (social, ego, development) we find fear yielding to pleasure as the primary emotion behind the motivation. As long as there have been sticks and carrots, managers have intuitively recognized the fear-pleasure dichotomy in their dealings with their people. The dichoto-

TABLE 8.1. Maslow's Hierarchy of Needs

Needs	Examples
Body	Oxygen, food, sex
Safety	Industrial safety, home security
Social	Sense of belonging, friendship
Ego	Recognition, praise
Development	Aspirations, striving for excellence

my has also been explicitly recognized in the management literature, and in the next section we'll consider two theories of McGregor (1960) and their implications for modern organizations and their leaders.

THEORY X AND THEORY Y

McGregor's (1960) seminal book, *The Human Side of Enterprise*, considers the management of organizations from the standpoint of human motivation, using Maslow's hierarchy as an important base model. He concludes that there are two fundamentally different types of managers who operate from two different sets of assumptions about human behavior.

The first type of manager operates from a set of assumptions that McGregor labels *theory X*. These managers believe that people

1. Dislike work
2. Are not trustworthy and need to be watched constantly
3. Cannot make good decisions without close supervision
4. Need prodding to complete even the simplest of tasks

A manager believing these things is naturally led to a suspicious, coercive management style, where careful monitoring and punishment are the rule. By contrast, what McGregor calls the *theory Y* manager believes that people

1. Find work enjoyable
2. Can be trusted to work even in the absence of supervision
3. Can make good decisions autonomously
4. Can finish tasks without constant prodding

Because of these beliefs, the theory Y manager is led to a more supportive, praise-oriented style of management.

It seems odd that two such disparate viewpoints can coexist. After all, managers aren't necessarily stupid, and they are likely to adapt over time in trying to improve. How could such different views coexist in the long run? I think the answer is that both strategies are *locally optimal*. A solution is "locally optimal" if small changes in decision variables about the current operating condition result in a degradation in performance; thus both theory X and theory Y managers find that there is little way to improve their performance in the neighborhood of their chosen solution.

For example, as theory X managers flog their people—as they exert their suspicious, manipulative, often punishing brand of management—organizational output improves to a point; the people under their control tend to react defensively, reluctantly, and sometimes suspiciously, thereby confirming the theory X viewpoint. On the other hand, theory Y managers, by trusting their people and delegating authority and responsibility, find increased output; thus their life view is also confirmed. This seems strange, that both types of manager are locally correct, but the nonlinearity of human behavior—the nonlinearity that is implicit in Maslow's hierarchy—is enough to permit multiple optima to exist.

At this point, we might be tempted to say that we simply have two different solutions, each with its own merits. But, as McGregor goes on to argue, theory Y is the better approach in the long run, because it appeals to higher motives by emphasizing ego gratification. This is significant, as once employees take responsibility for their lives and seek goals autonomously that enrich individuals and the organization together, the theory Y manager has an easier time keeping the enterprise running smoothly.

Contrast this to the situation faced by theory X managers. The need to keep things running requires increasing threats to maintain the same, if not diminished, levels of performance. Monitoring efforts are effective as long as theory X managers are standing there; as soon as they turn their backs, the troops look for ways to slow things down, gum up the works, or otherwise obstruct progress. Over time, the theory X coercive style requires an escalation of threats that are decreasingly effective.

All of the foregoing would just be very nice management theory, if that was as far as it went; but there are many war stories to be told about the positive changes in individuals and organizations that have occurred when management shifted from theory X to theory Y.

Case of the Sluggish Secretary

Consider the case of the sluggish secretary. There was once a secretary in a government agency who was reluctant to do any work not part of her normal routine of phone reception, filing, and light typing. Department co-workers who tried to approach her with out-of-the-ordinary jobs were met with evasion and avoidance, and many shared the view that she was less than competent.

At the time, the secretary worked for a boss who was quick to find fault and slow to forgive. One day, job assignments were switched, and she was assigned to work with an up-and-coming young manager. He, too, had had his share of bad experiences with her, but he was too busy to dwell on the past and immediately gave her responsibility to organize a seminar series. The job required that she perform many new tasks, including letter writing, mass mailing, phone calls, sales calls, and follow-up, as well as on-site registration and client contact. To everyone's surprise, this once-sluggish secretary attacked her new responsibilities with a vengeance, the seminar series was a success, and the woman's confidence and helpfulness spilled over into her other duties. When asked to do new things she no longer avoided or evaded them. She simply tried to dig in and do the job.

In looking back over this description, we can easily see what accounts for the change. She went from a theory X situation, with suspicion, punishment, and blame lurking at every turn, to a theory Y situation, with responsibility, trust, reward, and responsibility the watchwords of the day. The transformation that occurred can be cited repeatedly in stories of rapid change in outlook, attitude, and performance. The magic that theory Y worked in this case is not unusual. But can these lessons apply in a wider context? In the next section, we'll consider an organizational success story that started before the turn of the century.

Theory Y at the Lincoln Electric Company

One-on-one stories of theory Y success are somewhat reassuring, but is it possible, in a large organization, to have theory Y permeate corporate nooks and crannies from the farthest foreign office to the factory floor? The answer is yes, and we need not turn to Johnny-come-lately, high-tech miracles in San Jose, Tokyo, or anywhere else. Since the turn of the century, a manufacturer of arc-welding and industrial electric equipment in Cleveland, Ohio, has been satisfying customers, stockholders, and employees with its own home-grown formula for internally cooperative, externally competitive success.

Lincoln Electric Company was founded by engineer John C.

Lincoln in 1895; he was later joined by his brother James F. Lincoln, a skilled motivator, leader, and businessman. Over the years, the company developed the *incentive management system,* which combines belief in people, guaranteed employment, incentive-driven piecework (with payment based on quality and quantity), more-than-window-dressing employee participation in decision making, and a merit-driven bonus system. In every year since 1934, the company has paid an annual bonus that has averaged nearly 100 percent of each employee's regular annual pay. The system is detailed in a book by James F. Lincoln (1951) entitled *Incentive Management.*

Many of the features of the company's system resemble those of "new and enlightened" Japanese management systems. It is remarkable to think that a smokestack company in what has become the "rust belt" of the midwestern United States has been "enlightened" since before the turn of the century. This company continues to innovate, compete, and thrive because of its belief in its people—because of its theory Y culture. Later we will consider a number of organization-design decisions that help create such a culture. For now, consider one of your own organizational experiences in light of theories X and Y.

WHY FORM OR JOIN ORGANIZATIONS?

In this chapter, we've been examining organizations and how they're run. In a moment we'll consider some aspects of organizational design, but for entirely too long we've ignored a fundamental question: Why do we form, and join, organizations? At first, the question sounds almost silly. After all, forming organizations is what you do when you want to get organized, right? The question

---

Online Exercise 8.1

Consider an organization with which you are familiar. In a paragaph or two, explore whether it is a theory X or theory Y organization and explain your reasoning.

may seem a bit more sensible, however, if we acknowledge an alternative to the tightly knit grouping we call an organization.

In a market economy, many transactions are made through the marketplace; we'll call the combination of an entrepreneur and the group of individuals he or she engages through market transactions a *confederation*. To simplify matters, let's imagine a widget manufacturer organized as a two-person confederation, with an entrepreneurial boss and a subcontractor. The entrepreneur designs the widget, starts the company, lines up clients, sets up the initial manufacturing process, and performs the initial production runs. After a time, the boss gets tired of being a one-man band and turns to the market on a daily basis to engage the services of a widget subcontractor—possibly different subcontractors on different days—to help with the manufacturing of widgets.

Contrast this structure to that of an identically sized *organization*. In a two-person widget manufacturing organization, the entrepreneurial boss still exists and still performs the same duties, but now the widget helper is not hired on a daily basis but is an *employee*. The employee performs the same duties as the widget subcontractor, but instead of turning to the free market each day, the entrepreneur chooses to hire one person and to stick with that person for a longer period of time.

The confederation is something of an artificial construction, but the purpose here is to understand why organizations are formed by thinking of a suitable alternative. When we contrast the two structures, there don't appear to be many significant differences between them at first. Both structures make the same widgets; both structures have the same number of people in the same roles. The only difference is that the organization chooses not to turn to the free market as frequently as does the confederation to hire its help. (Nor does the employee choose to turn to the free market to look for work as often as does the subcontractor.) In other words, employers and employees agree, somewhat implicitly, to form a longer-term union. Recognition of this *mutual commitment* is very important and should be kept in mind in the design and leadership of organizations, but to better understand where it comes from, we need to ask what advantages organizations have over confederations. Such advantages must exist, or otherwise entrepreneurs would prefer to turn to the free market to engage helpers for their enterprises.

It is tempting to explain the advantages of an organization

directly. For example, perhaps the entrepreneur is motivated to hire a permanent employee—instead of turning to the marketplace to hire a subcontractor—by (1) cheaper wages, (2) better control, or (3) clearer identity. However, none of these reasons withstands much scrutiny. If the wages paid to a subcontractor are greater than those paid to an employee, they are greater because of the greater risk taken by a subcontractor facing uncertain duration of employment; in the long run the free market can be expected to evaluate this trade-off between risk and payment accurately. Control is no better an explanation of why organizations exist, because the entrepreneur exercises equally clear management authority in both an organization and in our hypothetical confederation. Moreover, because the identity of an organization—or a confederation—is based on the cost, reputation, and quality of what is produced, and because neither entity has an inherent advantage along these dimensions, identity cannot be used to explain why organizations are formed and joined.

If such direct effects fail to explain an organization's advantage, what does? Turning to the essential difference between the confederation and the organization—the confederation's more frequent use of the free market to engage services of individuals—helps shed some light on the organization's advantage. If an entrepreneur turns to the free market to hire subcontractors, chances are that different bidders will be engaged on different days. As a result, the confederation will be subject to a number of *costs* that a corresponding organization with its more-or-less permanent workforce need not pay:

1. Costs of hiring, paperwork, and contracting
2. Costs of training
3. Costs of incompetent hires

Frequent turnover of subcontractors in the confederation will require that whatever paperwork, legal fees, and other sign-up costs are necessary will be incurred each time a different subcontractor is engaged. Worse than this, each time a new subcontractor is hired, he or she must be made familiar with the local particulars of making widgets, and in most manufacturing processes training costs can be significant. Finally, when one turns to the marketplace, one can never be certain that a hire is going to work out. There are plenty of smooth talkers with good-looking vitas who turn out not to be very good widget makers. This analysis is presented from the entrepreneur's point of view, but the same conclusions are reached if it is approached from the employee's or subcontractor's perspective.

Taken together, these seemingly secondary factors, what economists call *transaction costs*, are why we form and join organizations (Coase, 1988). Although some of these costs are mundane things like paperwork and legal fees, many of them are *costs of knowledge* (Sowell, 1980). Confidence in the competence of an employee (or an employer) does not come free of charge; it is knowledge that is gained over a period of time. (And when an employee does not work out, it can be knowledge that is very expensive indeed.) As a result, when employers find employees they are happy with, and when employees find work situations they are happy with, they tend to stick with each other. In other words, they form an organization, and in so doing they make an implied commitment to one another. It is not a commitment for life necessarily, but it is a commitment not to consult the free market as often as they might if knowledge came more cheaply than it does.

THE ORGANIZATION'S DILEMMA— OR WHY COOPERATION AIN'T EASY

The previous section suggested that organizations are formed because of transaction costs and that implicit in the act of joining an organization is the act of mutual commitment between employer and employee. But not continually turning to the marketplace for employees (or employers), organization members come together and work toward organizational goals. Within the confines of an organization, however, there can be large variations in the degree of commitment that exists. In a sense, the mutual commitment implicit in the forming or joining an organization is the base level of cooperation that must exist. Whether this minimal level turns into large-scale cooperation and trust depends greatly on the behavior of the members of the organization. The health of the enterprise can depend on the formation of this higher level of cooperation, because many organization difficulties such as labor strife, poor product quality, or bad customer service all have their roots in a lack of cooperation between workers and management. In this section we'll consider the root problem of cooperation in somewhat abstract form through the *prisoner's dilemma problem*. We will find why management actions that seem unrelated to the workforce can spill over and degrade the cooperative spirit that is the basis of organizational activity.

The prisoner's dilemma problem is a model of the essential conflict that arises in many settings that call for cooperation. Suppose that a group of individuals is arrested for one criminal act. In such cases, it is usually in the group's interest for each individual to refuse to speak to authorities (in other words, to cooperate with each other), but the authorities may offer leniency in exchange for one individual's testimony against the group. A payoff structure that tempts the individual at the expense of the group is common in many cooperative situations. To connect this to our subject, let's rename our version of the prisoner's dilemma problem the *organization's dilemma*, and fill in some details of the conflict.

Consider the boss (A) and employee (B) from the widget organization of the previous section, and imagine that each has a fundamental, binary decision to make: to *cooperate* or *defect*. For A, the choice might be whether to give B a larger bonus (cooperate) or to give himself a golden parachute, perks, and a larger salary (defect). For B, the choices might be whether to work more diligently (cooperate) or to loaf (defect). The relative payoff matrix is shown in Table 8.2. Here the payoff is represented as an ordered pair, with A's payoff listed first and B's next. Looking at the table, we see that, if both cooperate, a good time is had by all. If one or the other (but not both) defects, that individual gets a reward while the other individual (the sucker) receives a fairly severe punishment. If the sucker now defects as well, that individual's payoff situation improves somewhat, but the first defector loses all previous gain and then some.

Although this game matrix is relatively simple, it raises a number of interesting questions, perhaps the most important of which is how cooperative behavior ever evolves under such circumstances (Axelrod, 1984). For our purposes, we see how almost all self-serv-

TABLE 8.2. The Organization's Dilemma Problem

A/B Actions	B Is Diligent	B Loafs
A gives bonus to B	(good, good)	(worse, better)
A gives perks to A	(better, worse)	(bad, bad)

ing behavior in an organizational setting can be perceived as defection from proper cooperation. This should make it clear that although cooperation has the potential to maximize an organization's output, the challenge of sustaining cooperation over the long haul makes it difficult to realize and maintain.

For just such reasons, it is important to design organizations from the very start with a view toward mutual commitment and cooperation. With that in mind, we'll consider a few key issues of organizational design in the next section.

SOME ASPECTS OF ORGANIZATIONAL DESIGN

In this section, we consider several important guidelines of organizational design:

1. Place decision making at the point of attack.
2. Create flat organizational and salary structures.
3. Eliminate or minimize perks.
4. Choose measures that matter.

We will briefly consider each of these guidelines in somewhat more detail.

Place Decision Making at the Point of Attack

The theory Y organization, believing in the capability of its members, tries to place authority and responsibility for decisions and their consequences at the point of attack. For example, decisions about modest expenditures are made by those who have the need in question. It has always puzzled me why organizations bother to hire engineers at high salaries and benefits without giving them any budget authority to do their jobs. There are organizations where approval for a $200 piece of software may require four or five signatures; in such situations, whatever frugality is achieved by those signatures is not worth the cost in resentment they engender.

Of course, this principle does not apply only to small decisions. As an example, hiring—one of the most important decisions made in organizations—should occur close to the point of attack, as should many important decisions. The benefit of moving important decisions downward is clear. When decision-making authority is held by the powers on high, there is a tendency for the lower level to take their role in the decision lightly. And who can blame them? Many times in the past, their recommendations may have been over-ridden, and over time they simply learned to float trial balloons without much reflection. When the decision is completely delegated, the lower level rises to the task, spending more time than the higher level would in weighing and evaluating the decision, as well they should. After all, the decision affects them most directly, and they are the ones who must live with its consequences.

Create Flat Organizational and Salary Structures

Theory Y organizations tend to be fairly lean and flat. Because much of the authority is delegated to where it will do the most good, there is lower management overhead; thus an individual manager can have a larger number of workers reporting to him or her (larger span of control). Moreover, many staff functions become subsumed in the line function. The most ponderous hierarchies form in staff organizations having little responsibility to the real world.

Almost to prove this point, I recently received a glossy color brochure published for an in-house administrative unit within my university. The slick brochure went on for pages, showing pictures and descriptions of the many managers, assistants, assistant to assistants, consultants, and *their* assistants. The brochure never mentioned what it was that any of these people did, or why (or why it was necessary to waste taxpayers' money on this self-trumpeting brochure). All staff organizations can take on a life of their own like this. Fortunately, for-profit companies have some motivation to trim such layers of fat; but, even then, internal staff organizations have no natural predators, and unless they are carefully monitored by a watchful management they can quickly consume ever-larger pieces of the organizational pie, much larger than their function might otherwise deserve.

Contracting externally for such services can make sense; sometimes centrally required services can be distributed to line units. Townsend (1970), for example, recommends the elimination of all personnel and purchasing departments, replacing them with hiring and buying done at the point of attack. Personnel and purchasing departments are often justified on the basis of economy of scale, and such savings may exist at first, but then the staff monster takes on a life of its own that overshadows the original economy.

A flat structure has the additional advantage that salary ratios, top to bottom, can be better controlled. It is easier to justify jillion-dollar executive salaries in 20-layer firms than it is in flatter organizations. This is no small matter, but recent concerns about executive pay expressed in the mass media miss the point that high-output organizations are necessarily cooperative organizations. As we saw in the two previous sections, organizations are constructed from a kind of implied mutual commitment between their members, and cooperation grows—when it does—on top of this initial commitment. When the CEO takes out a golden parachute or permits executive pay or bonuses to be raised in hard times, this is a defection from cooperative behavior that can halt, and possibly destroy, the movement toward the cooperation that is so necessary to healthy organizations.

Eliminate or Minimize Perks

In many organizations, high-level managers have fancy offices, are driven about in limousines, and are flown from place to place in corporate jets. In theory Y organizations, where the idea is to engender the pursuit of group goals, an individual feeding at the perquisite pot represents a defection from the organizational mutual commitment and cooperation.

Much is made of the lack of perquisites in some Japanese corporations; the spirit of cooperation this helps create should not be underestimated. This was driven home to me on a recent trip to Japan, when I visited Mitsubishi Electric near Osaka and was shown through a number of research labs. I walked with my hosts, a group of Mitsubishi research engineers and managers, past the laboratory director's office; by all signs it was not much different from a salaried employee's office. No gilded quarters for the head honcho, no bevy of secretaries, just a simple office like that of ordinary workers. Perhaps the most revealing detail was the pride my hosts took in telling me about their big boss who had an office just like theirs.

Choose Measures That Matter

In designing organizations, it is important to choose appropriate measures of performance. Much has been made of the Japanese tendency to watch market share (a long-term measure of success) and the American tendency to watch quarterly return on investment (a short-term measure). The reasons for these choices are complex and not simply explained (or changed). Nonetheless, it should come as no surprise that managers who are judged by their company's performance as measured by market share will naturally take a longer view than managers who are graded on company quarterly return on investment.

Another example may be found in the modern university, where tenure and promotion decisions are often (partially or largely) based on quantitative measures of research productivity (research funding obtained, numbers of papers published). Again, the reasons for these choices are complex, but we should not be surprised when faculty take the measures of their performance quite seriously and spend more time proposing and writing than preparing for class.

The point of these examples is not to point a finger or complain about lousy managers or college professors (although such carping has become grand sport in the popular media). Quite the contrary. The point here is to suggest that when things go awry in organizations, it is often because the measures of performance and the incentives that go along with them have created exactly the mess that exists. Although, in the short term, we can be fairly dense and slow to change, in the long haul, human beings are incredibly adaptive to the most subtle nuance and detail. Numbers—especially those connected to incentives or disincentives—take on a life of their own, and therefore it is important to question the regular publication of *any* statistic. If a prior analysis suggests that employee adaptation to a particular measure might skew organizational activity badly, then dissemination of that number is a risky business at best.

Most organizations have multiple objectives, and if measures for some of the objectives don't exist or are less reliable than others, those objectives will be taken less seriously. The flip side of this coin is that choosing measures wisely can positively affect both group and individual performance. A key feature of the Lincoln Electric system and of Scanlon plans (Lesieur, 1958), in general, is the association of pay with performance, and both systems spend time making sure that the arrangement is fair to employees, management, and stockholders. Ad hoc attempts to adjust bonus or incentive systems (so they enrich management or stockholders at the expense of employees) are usually fairly transparent and will be perceived as the defection from mutual commitment and cooperation that they are.

LEADERSHIP

Good leadership is easy to recognize but hard to describe. In this section, we will consider a number of dos and don'ts of leadership:

1. Praise the praiseworthy.
2. Be cautious in criticism.
3. Lead, don't dump.
4. Talk, don't write.
5. Prefer the simple to the complex.
6. Set objectives and expectations.
7. Have a vision.

We'll examine each of these dicta in somewhat more detail.

Praise the Praiseworthy

We extolled the power of praise at length in Chap. 4, but it is such an important habit that it bears repeating. We all love praise, but we are remarkably stingy in giving it out. As we noted, praise should be truthful and specific to be effective. A related idea recognizes that organizations work in teams and that praise needs to be spread around to everyone who plays a role. If there is doubt about who deserves credit, a praising should tend to be inclusive rather than exclusive. Moreover, if there are individuals who have risen above the rest, their special efforts should be acknowledged beyond those of the group.

Be Cautious in Criticism

This is also a topic we discussed in Chap. 4, but it, too, bears repetition. As difficult as it is for us to praise, it is ridiculously easy for us to criticize. This is especially remarkable because criticism is usually so utterly ineffective. Human beings have an almost inexhaustible capacity to shield themselves from the harshest criticism or punishment. Therefore, when criticism is necessary, it is important to make sure that the individual being criticized knows that you are in his or her corner, that you would simply like to see a change in behavior. In other words, if we focus on behavior as separate from personality, and let the person know that we dislike the behavior but still respect the individual, there is some possibility that skillfully applied criticism can break through and invoke changes to the good.

Lead, Don't Dump

Hundreds of management books have intoned the three-word management mantra, "delegate, delegate, delegate," yet thousands of managers have misinterpreted those words to mean that they should collect the 10 worst chores sitting in their in-basket and dump them on their workers. These managers are 180 degrees out of phase. Managers should take the most important tasks and give them to their people and do the bottom-feeder stuff themselves.

My first department head at the University of Alabama, William D. Jordan, was a master at keeping important material in our in-baskets and keeping nitpicky stuff out. Universities have an enormous capacity to make ridiculous requests for reports and trivia, and Bill Jordan was as good as anyone I know at handling such matters with minimal intervention from us. In working with my own students, I try to shield them from the administrivia of contracts and universities and keep them happily plugging away at the fun, technical research. I don't know that I succeed as well as Bill Jordan did, but I always try to hold to his example as the way it should be done.

Talk, Don't Write

I have lived in organizations where the first response to even the simplest request for a decision was "Put it on paper." This is a wonderful tactic if you are interested in bringing progress to a screeching halt, but most normal business can in fact be handled informally, via one-on-one discussions, in a matter of minutes—and that is the way it should be. I have too often seen problem situations where the first response of the battling parties is to shoot memos around the organization. This is a telltale sign of a theory X organization and is almost always counterproductive. Interpersonal conflicts are an occasion for people to be *talking*, not writing. Putting bad news on a piece of paper allows the bad news to live in infamy forever, and it erects a barrier between you and your adversary, because it does not permit that person to respond immediately.

A corollary to "No bad news on paper" is that it can be particularly effective to praise people on paper, because paper is more permanent and carries greater weight than the spoken word.

Prefer the Simple to the Complex

Engineering training encourages us to build elaborate models of physical processes, and the complexity of modern technology often forces us to think in terms of steeply hierarchical and interconnected systems. When it comes to our organizational interactions, however, we are better off *keeping things simple*. When someone comes up with a new form or survey, someone else should ask whether it is really necessary. When yet another committee designs yet another rule, someone should ask if it is really needed, and why. In setting forth goals, it is important they be stated simply, which leads us to our next topic.

Set Objectives and Expectations

Setting and resetting objectives should be a regular part of organizational behavior. A useful little book by Blanchard and Johnson (1982), *The One-Minute Manager,* suggests having workers prepare one-page sheets for each key objective, thereby committing key goals to paper *simply*. Some organizations have enshrined goal setting in a symbolic, *managing by objectives* ritual. Such exercises are more smoke than fire. I recall that at one organization where I worked we were required to devise a five-year plan annually; every year we held a ritualistic "planning" meeting, where it seemed that we did little more than add one year and 10 percent.

Real planning and setting of objectives are too important to ritualize. Moreover, they are dynamic and constantly required. In short, these are activities at the core of what organizations must do if they want to accomplish anything.

Beyond goal setting, it is up to a leader to set expectations of conduct and behavior within an organization. This occurs, first and foremost, with the leader's own conduct. It is unreasonable for a stubborn despot to expect cooperative, conflict-free human relations within his organization. Setting a consistent good example is the place to start, but the leader's responsibility goes beyond good example. He or she must also try to articulate a vision of appropriate organizational behavior. This vision can be propagated by formal training, by organizational culture, or by both, but it is important that it be propagated.

Have a Vision

Perhaps the most important thing a leader can do for an organization is to have and communicate a vision of the future. It is difficult to say exactly how one comes to such a vision, but we can identify some of its elements:

1. It articulates the organizational mission.
2. It plays to the organization's strengths.
3. It overcomes or diminishes the organization's weaknesses.
4. It identifies critical new opportunities.

The base requirement of vision is to understand your organization's mission and to articulate it simply. Conciseness requires a tight editing and clears away clutter. In addition, conciseness makes it easier to carry the message to others and have them remember, understand, and accept the mission.

With a carefully crafted mission statement, it is possible to examine your organization's activities and determine whether they are consistent with that mission. In the case of activities the organization is good at, one strategy is to try to do them more and better. In nature, this exaggeration of a good feature—the evolutionary transformation of a bone into a tusk, for example—is called *hypertrophy*. In business it's only good sense to play to one's strengths.

Weaknesses are another matter. A perceived weakness that can be ignored without much harm operationally isn't much of a weakness. On the other hand, there are occasions when a weakness prevents or slows an organization from achieving its mission. In these cases the weakness must be attacked. Sometimes, in a larger organization, this calls for hiring people who can make up for the weakness. In a smaller organization it may mean training existing people to improve in the area of weakness.

The last part of having a vision is the most difficult: identifying new opportunities. Something is "new" because it didn't previously exist; so, by definition, this part of having a vision is the most creative. Coming up with new ways to carry out an organization's mission is critical because most organizations live in a competitive

---

Online Exercise 8.2

Consider an organization with which you are familiar. In a brief essay, prepare a mission statement for this organization, then identify its strengths and weaknesses and its key opportunities.

---

environment, and one's competitors do not sit still. Keeping an eye on one's competitors is useful; so-called *benchmarking* formalizes this activity. By thinking only about the present is a sucker's bet. Because of development time, if you simply chase your competitors, you will always be a step behind and a day late. Therefore defining and going after new opportunities is absolutely imperative in maintaining long-term organizational health. The structured-brainstorming method outlined in the previous chapter can be useful in trying to accomplish this, and fostering a nurturing environment where new ideas are welcomed and rewarded also can be very helpful. However it is done, new opportunities must be identified and pursued, thereby renewing the organization and keeping its mission alive.

Summary

In this chapter we have examined organizations and their leadership. Our primary observation has been that the others'-eyes principle of human relations can go a long way toward promoting better organizational design and leadership. Specifically, we have tried to generalize the one-on-one lesson by looking at Maslow's hierarchy of needs and its application in McGregor's theory X and theory Y models. Although locally optimal results can be achieved by theory X managers who distrust people and use punishment and fear to drive their organizations, we have seen how praise and pleasure can be used more effectively in the hands of skilled theory Y managers.

But organizations are more than one-on-one relationships that have been scaled up. An examination of some current theory in economics suggests that we form and join organizations largely to avoid the continuing transaction costs of the market, costs that can be substantial, particularly those associated with making assessments about the suitability of an employee (or employer). The effect of these substantial costs is to make it desirable and profitable for both employers and employees to make a mutual commitment to one another. This mutual commitment—born of self-interest—can lead to a higher form of cooperation; however, through the examination of the organization's (prisoner's) dilemma problem, we have seen why cooperation is so difficult to achieve. If higher levels of cooperation are to be achieved, the organization's dilemma suggests that both managers and employees must be sensitive to defections from cooperative behavior, especially in actions that might be perceived by others as self-serving.

These ideas have led us to consider their application in organizational design and in leadership. Organizations should be designed for results, not appearances. Many of the bad habits and practices that develop in organizations are the result of the desire for symbolic aggrandizement of their managers. It is more important to focus on real-world results and output; when this is done, it is interesting to see how the perquisites, inflated salaries, empires, and other bad signs disappear.

In the realm of leadership, it is important to deal with people humanely, according to good, one-on-one principles of human relations; but it is also important to have a vision of where the organization must go. This requires a sense of mission, a knowledge of strengths and weaknesses, and an ability to discover new opportunities. These challenges are as difficult as the most demanding of technical tasks, but they are challenges that we as engineers must tackle if our organizations are not to become second-rate producers of second-rate goods.

Offline Exercises

1. Consider an organization in which you have worked. Make a list of the ways in which the organization hindered you in the performance of your job.
2. Consider a manager you have worked for. Make a list of things he or she did that you would classify as theory X behavior and make another list of things he or she did that you would classify as theory Y behavior.
3. Consider a leader of historical renown. Write a brief essay regarding his or her accomplishments in the light of theory X and theory Y.
4. Scan a newspaper for evidence of corporate or organizational difficulty. Write a short essay analyzing the situation in the light of theories X and

Y. Read between the lines when necessary to make your analysis, but be sure to separate fact from speculation in your writing.

For Further Reading

AXELROD, R. M. (1984). *The evolution of cooperation.* New York: Basic Books.

BLANCHARD, K., AND S. JOHNSON (1982). *The one-minute manager.* New York: Berkley.

COASE, R. H. (1988). *The firm, the market, and the law.* Chicago: University of Chicago Press.

LESIEUR, F. G. (ed.) (1958). *The Scanlon plan.* Cambridge, Mass.: MIT Press.

LINCOLN, J. F. (1951). *Incentive management.* Cleveland, Ohio: Lincoln Arc Welding Association.

McGREGOR, D. (1960). *The human side of enterprise.* New York: McGraw-Hill.

MASLOW, A. (1970). *Motivation and personality* (2d ed.). New York: Harper & Row.

SAFIRE, W., AND L. SAFIR (eds.) (1990). *Leadership.* New York: Simon and Schuster.

SOWELL, T. (1980). *Knowledge and decisions.* New York: Basic Books.

TOWNSEND, R. (1970). *Up the organization.* New York: Knopf.

CHAPTER 9

Technical Skills + Life Skills = Engineer

Eight chapters ago we started with the idea that success as an engineer is determined by a combination of life skills and technical skills and that engineering education has traditionally focused almost entirely on the latter. Hoping to partially remedy this flaw, we have given the intervening pages to briefly surveying those key life skills that an engineer needs to be most effective. From communication skills such as writing and presenting, to human relations, to the management of time and money, we have hit the highlights of those individual skills that will have the most impact on an engineer's daily routine. We have also discussed group skills, from structured brainstorming to the design and leadership of organizations, that can help you be more effective as you start to play a leadership role in your organization.

The usual reaction to this material comes in two flavors: so what and anxiety. The so-what camp seems to take the engineer's need for mastery of both technical and business skills for granted and is comfortable with the near-schizophrenia required by a profession with such broad intellectual and emotional demands. The more common reaction is that of anxiety. Throughout much of its modern history, our profession has been a battleground to win control of the business or the technical side of the engineer's mind. (For example, attempts to unionize, enforce universal licensing, and wrest control from business management in the name of "scientific management" have all been made and have all failed.) This tension has played a central role in the formation of our profession and of society's reaction to it (Layton, 1971).

Intellectually, engineers are in the prescription business; we decide what should be done and how things should change, and this requires that we become intimately involved in the operation and leadership of the organization that do the doing and changing. In other words, any search by engineers for professional or business purity is doomed to fail, because it is the *combination* of technical skill and the ability to get the job done that gives engineers their competitive advantage over those interested only in technical matters (scientists) or business matters (businesspeople) alone.

Unfortunately, in an age of increased specialization, the need to balance such different proficiencies gives way to a pressure to break apart the hybrid, to replace the engineer as technician-businessperson with specialists, that is, technicians and businesspeople. This is a double mistake. It leads to businesspeople who rely too heavily on a financial, marketing, and legal bag of tricks that in the long run has little to do with the organization's fundamental business of innovation. In addition, overly technical technicians become enamored with modeling for its own sake, underestimate the importance of the bottom line, and continually miss the key point that products (and services) are designed, built, bought, and used by human beings, complex creatures with nonlinear motivations.

The hope of this book is that engineers who are better able to play their dual technical-business role can bring to their organizations a passion for products and services *and* for the technology that drives them. This is a tall order, and it will require an increased willingness to get involved in the messy affairs that necessarily accompany the accomplishment of anything by a group of human beings. At times, the ambiguity will exceed what we find sketching with design pencil in hand, typing or mousing around in front of our workstations, or punching the buttons of our calculators; and at times, it will be uncertain whether our efforts are having any impact. But these efforts are necessary if our organizations are to succeed in the necessarily passionate and innovative struggle for survival now and into the next century.

For Further Reading

LAYTON, E. T. (1971). *The revolt of the engineers: Social responsibility and the American Engineering profession.* Cleveland, Ohio: Press of Case Western Reserve University.

Index